Science, Technology and Medicine in Modern History

General Editor: John V. Pickstone, Centre for the History of Science, Technology and Medicine, University of Manchester, England (www.man.ac.uk/CHSTM)

One purpose of historical writing is to illuminate the present. At the start of the third millennium, science, technology and medicine are enormously important, yet their development is little studied.

The reasons for this failure are as obvious as they are regrettable. Education in many countries, not least in Britain, draws deep divisions between the sciences and the humanities. Men and women who have been trained in science have too often been trained away from history, or from any sustained reflection on how societies work. Those educated in historical or social studies have usually learned so little of science that they remain thereafter suspicious, overawed, or both.

Such a diagnosis is by no means novel, nor is it particularly original to suggest that good historical studies of science may be peculiarly important for understanding our present. Indeed this series could be seen as extending research undertaken over the last half-century. But much of that work has treated science, technology and medicine separately; this series aims to draw them together, partly because the three activities have become ever more intertwined. This breadth of focus and the stress on the relationships of knowledge and practice are particularly appropriate in a series which will concentrate on modern history and on industrial societies. Furthermore, while much of the existing historical scholarship is on American topics, this series aims to be international, encouraging studies on European material. The intention is to present science, technology and medicine as aspects of modern culture, analysing their economic, social and political aspects, but not neglecting the expert content which tends to distance them from other aspects of history. The books will investigate the uses and consequences of technical knowledge, and how it was shaped within particular economic, social and political structures.

Such analyses should contribute to discussions of present dilemmas and to assessments of policy. 'Science' no longer appears to us as a triumphant agent of Enlightenment, breaking the shackles of tradition, enabling command over nature. But neither is it to be seen as merely oppressive and dangerous. Judgement requires information and careful analysis, just as intelligent policy-making requires a community of discourse between men and women trained in technical specialities and those who are not.

This series is intended to supply analysis and to stimulate debate. Opinions will vary between authors; we claim only that the books are based on searching historical study of topics which are important, not least because they cut across conventional academic boundaries. They should appeal not just to historians, nor just to scientists, engineers and doctors, but to all who share the view that science, technology and medicine are far too important to be left out of history.

Titles include:

Julie Anderson, Francis Neary and John V. Pickstone
SURGEONS, MANUFACTURERS AND PATIENTS
A Transatlantic History of Total Hip Replacement

Roberta E. Bivins
ACUPUNCTURE, EXPERTISE AND CROSS-CULTURAL MEDICINE

Linda Bryder
WOMEN'S BODIES AND MEDICAL SCIENCE
An Inquiry into Cervical Cancer

Roger Cooter
SURGERY AND SOCIETY IN PEACE AND WAR
Orthopaedics and the Organization of Modern Medicine, 1880–1948

Jean-Paul Gaudillière and Ilana Löwy (editors)
THE INVISIBLE INDUSTRIALIST
Manufacture and the Construction of Scientific Knowledge

Christoph Gradmann and Jonathan Simon (*editors*)
EVALUATING AND STANDARDIZING THERAPEUTIC AGENTS, 1890–1950

Alex Mold and Virginia Berridge
VOLUNTARY ACTION AND ILLEGAL DRUGS
Health and Society in Britain since the 1960s

Ayesha Nathoo
HEARTS EXPOSED
Transplants and the Media in 1960s Britain

Neil Pemberton and Michael Worboys
MAD DOGS AND ENGLISHMEN
Rabies in Britain, 1830–2000

Cay-Rüdiger Prüll, Andreas-Holger Maehle and Robert Francis Halliwell
A SHORT HISTORY OF THE DRUG RECEPTOR CONCEPT

Thomas Schlich
SURGERY, SCIENCE AND INDUSTRY
A Revolution in Fracture Care, 1950s–1990s

Eve Seguin (*editor*)
INFECTIOUS PROCESSES
Knowledge, Discourse and the Politics of Prions

Crosbie Smith and Jon Agar (*editors*)
MAKING SPACE FOR SCIENCE
Territorial Themes in the Shaping of Knowledge

Stephanie J. Snow
OPERATIONS WITHOUT PAIN
The Practice and Science of Anaesthesia in Victorian Britain

Carsten Timmermann and Julie Anderson (*editors*)
DEVICES AND DESIGNS
Medical Technologies in Historical Perspective

Duncan Wilson
TISSUE CULTURE IN SCIENCE AND SOCIETY
The Public Life of a Biological Technique in Twentieth Century Britain

Science, Technology and Medicine in Modern History
Series Standing Order ISBN 978–0–333–71492–8 (hardcover)
Series Standing Order ISBN 978–0–333–80340–0 (paperback)
(*outside North America only*)

You can receive future titles in this series as they are published by placing a standing order. Please contact your bookseller or, in case of difficulty, write to us at the address below with your name and address, the title of the series and one of the ISBNs quoted above.

Customer Services Department, Macmillan Distribution Ltd, Houndmills, Basingstoke, Hampshire RG21 6XS, England

Tissue Culture in Science and Society

The Public Life of a Biological Technique in Twentieth Century Britain

Duncan Wilson
Wellcome Trust Research Associate, University of Manchester, UK

First published 2011 by
PALGRAVE MACMILLAN

Palgrave Macmillan in the UK is an imprint of Macmillan Publishers Limited, registered in England, company number 785998, of Houndmills, Basingstoke, Hampshire RG21 6XS.

Palgrave Macmillan in the US is a division of St Martin's Press LLC, 175 Fifth Avenue, New York, NY 10010.

Palgrave Macmillan is the global academic imprint of the above companies and has companies and representatives throughout the world.

Palgrave® and Macmillan® are registered trademarks in the United States, the United Kingdom, Europe and other countries

ISBN 978-0-230-28427-2   hardback

This book is printed on paper suitable for recycling and made from fully managed and sustained forest sources. Logging, pulping and manufacturing processes are expected to conform to the environmental regulations of the country of origin.

A catalogue record for this book is available from the British Library.

Library of Congress Cataloging-in-Publication Data

Wilson, Duncan, Dr.
    Tissue culture in science and society : the public life of a biological technique in twentieth century Britain / Duncan Wilson.
        p. cm.
    Includes index.
    Summary: "This book charts the social and cultural history of the scientific technique known as "tissue culture." It shows how tissue culture was a regular public presence in twentieth-century Britain, and argues that history can contribute to current debates surrounding research on human and animal tissue" – Provided by publisher.
    ISBN 978–0–230–28427–2 (hardback)
      1. Tissue culture–Great Britain–History–20th century. 2. Cell culture–Great Britain–History–20th century. 3. Embryology, Human–Great Britain–History–20th century. 4. Embryology, Experimental–History–20th century. I. Title.

QH585.2.W55 2011
571.5'3809410904–dc23   2011012466

10   9   8   7   6   5   4   3   2   1
20   19   18   17   16   15   14   13   12   11

Contents

List of Figures

List of Abbreviations

BBC	British Broadcasting Corporation
BMA	British Medical Association
BSE	Bovine spongiform encephalopathy ('mad cow disease')
BUAV	British Union for Abolition of Vivisection
CND	Campaign for Nuclear Disarmament
FRAME	Fund for the Replacement of Animals in Medical Experiments
GM	Genetically modified
MP	Member of Parliament
MRC	Medical Research Council
NAVS	National Anti-Vivisectionist Society
NIH	National Institutes of Health
UCSD	University of California, San Diego
UFAW	Universities Federation for Animal Welfare
UKCCCR	United Kingdom Co-ordinating Committee on Cancer Research

Acknowledgements

I am indebted to everyone who helped me research and write this book. My archival research depended in large part on the expertise and goodwill of staff at several libraries. Thanks, in no particular order, to the staff from Archives and Manuscripts at the Wellcome Library for the History of Medicine, the Science Fiction Archive at the University of Liverpool, the Video Archive at Oxford Brookes University, Archives and Manuscripts at the University of Cambridge, and the BBC Written Archives Centre. I am also grateful to staff at the British Film Institute and the BBC recorded archive, who provided me with important images and videos.

The Centre for the History of Science, Technology and Medicine (CHSTM), at the University of Manchester, was the perfect environment for this project. I am grateful to past and present CHSTM colleagues for their help and comments on various drafts. Thanks, then, to Fay Bound Alberti, Sam Alberti, Michael Brown, Katherine Foxhall, Jonathan Harwood, Vanessa Heggie, Emma Jones, David Kirby, Rob Kirk, Gillian Mawson, Chris Neilson, Neil Pemberton, Ed Ramsden, Carsten Timmerman, Elizabeth Toon and Michael Worboys. I owe special thanks to John Pickstone, who helped this project take shape and advised me throughout. It has been a great pleasure to work with John, and I look forward to doing so again in future.

I also owe thanks to historians outside Manchester, for their feedback on seminar papers and various ideas. Thanks to staff from the department of History at the University of Liverpool, the Centre for Medical History at University College, London, and Egenis, at the University of Exeter. I am grateful to Imogen Goold, from the University of Oxford, for her advice on the legal issues surrounding human tissue. And I owe particular thanks to Hannah Landecker, from the University of California, Los Angeles, who was happy to comment on various ideas and whose own work on tissue culture in the United States provided the inspiration for a British history. The advice from all these colleagues has been invaluable; any mistakes or omissions are mine alone. I also want to thank Michael Strang and Ruth Ireland at Palgrave, for their help in getting this book published. Last, but by no means least, I thank the Wellcome Trust for their financial support.

I also owe thanks on a number of personal levels, especially to my aunt, Elizabeth Clarke, and to Jean and Daish Sharma, who provided accommodation and hospitality during research trips to London.

I dedicate this book to my parents, John and Helen Wilson, who have always been supportive and never once told me to 'get a real job'. And to Nisha Sharma, who was a constant source of love, support and encouragement throughout this project.

1
Introduction

On 4 February 1935, Honor Fell, the director of the Strangeways Research Laboratory in Cambridge, wrote to the eminent physiologist Sir Henry Dale regarding an unscheduled visit from a newspaper journalist. The reporter from the *Sunday Dispatch* had called on the Strangeways laboratory to see the spectacular research that he, and many others, believed was taking place there. Unperturbed by Fell's refusal to speak with him, he nevertheless composed what she described to Dale as 'the most fantastic story'.[1] Dale replied the next day, urging Fell to simply ignore the press interest in her research. 'You will find', he claimed, 'that they soon get tired of it, and the trouble dies down'.[2]

Though Fell heeded Dale's advice, the 'trouble' did anything but die down. The Strangeways laboratory had been the subject of popular coverage for years, and would be for decades to come. The technique that Fell and her colleagues used in their research had already inspired several fictional accounts, radio talks and even a film that was shown at 10 Downing Street. These popular sources generally dwelt on the technique's 'stupendous possibilities'.[3] As Fell told Dale, the *Dispatch* believed that Strangeways researchers were '(a) on the point of creating life and (b) about to grow babies in bottles'.[4]

The technique in question is known as 'tissue culture' and involves the maintenance of excised human or animal tissues, cells and organs under laboratory conditions – or '*in vitro*'. It was first employed in 1907, when the Yale embryologist Ross Harrison sought to settle a dispute surrounding nerve growth by dissecting nerve cells from a frog embryo and culturing them in a suspension of lymph held in a sealed container. Culturing tissues, cells and even whole organs *in vitro* gave biologists an unprecedented ability to observe and manipulate living material apart from the body. As the twentieth century progressed, and especially after

1

World War II, tissue culture became a ubiquitous biological technique, underpinning high-profile work in embryology, biotechnology, cell genetics and virology.[5] Some scientists identify tissue culture as the 'greatest technical advance in medical science since the compound microscope' and list it as one of the ten greatest medical innovations.[6]

Historians claim that tissue culture had a profound impact on twentieth century biology: transforming cells and tissues from analytical entities to highly manipulable technologies, which could live apart from the body and be stored, used and exchanged like any other experimental artefact.[7] These histories generally look at tissue culture's impact in the United States; but it was just as central to biology in Britain, where scientists used the technique in pioneering work on cell fusion and *in vitro* fertilization. This book charts the British history of tissue culture, showing how it underpinned the work of particular scientists and institutions during the twentieth century. In addition to Honor Fell, we encounter Thomas Strangeways, who opened the first dedicated tissue culture laboratory in the 1920s and was hailed by *The Times* as 'the master of investigation in living cells'; Ronald Canti, who worked with Strangeways and Fell, and made cinematic films of cultured tissues; and Henry Harris and John Watkins, who were the first scientists to cross species barriers *in vitro* when they fused human and animal cells in the 1960s. These scientists are largely absent from existing histories of biology or medicine in Britain, but they were influential figures whose story is worth telling.

This book also offers a historical perspective on popular interest in the scientific use of tissue and cells. This is increasingly noteworthy given ongoing interest in stem cells, cross-species embryos and biobanks, as well as the recent controversy surrounding the retention of human tissues and organs in British hospitals. Motivated by these issues, a growing number of ethicists, anthropologists, sociologists and historians have turned their attention to medical and scientific research on bodily materials.[8] These writers tend to argue that popular attitudes to such research were, and are, overwhelmingly negative – highlighting a longstanding 'divide between scientific and social views of the body', where the public sees tissues and cells as sites of personal identity, kinship and experience, and researchers treat them as little more than an experimental resource.[9] Past examples of controversy, such as nineteenth century grave-robbing scandals, are often cited to support the claim that popular resistance has deep historical roots and changed little over time.

But I show that popular attitudes to tissue culture were enormously varied and changed substantially over the twentieth century. What is more, they reflected broader cultural concerns: e.g., the modernist fascination with the reconfiguration of lifespan and the body during the early twentieth century; arguments for and against the eugenic control of reproduction in the 1920s and 1930s; enthusiasm for medical 'magic bullets' in the 1950s; critiques of biology's inherent risks and social dangers in the 1960s; the rise of 'animal rights' in the 1970s; and the growing emphasis on patient autonomy during the 1980s and 1990s. I show, moreover, that these debates were historically specific and cannot be confounded across time.

Each chapter investigates the sets of activities within which tissue culture was constructed, used and given meaning, and shows how this involved a dynamic *engagement* between scientific practices and broader socio-cultural concerns.[10] Far from operating against the norms of popular actors and audiences, the scientists we encounter throughout this book drew upon and influenced them. From the outset, research on tissue culture was culturally mediated, and scientists popularized their work in line with cultural concerns. For example, our *Sunday Dispatch* reporter only troubled Honor Fell after several prominent scientists had claimed that tissue culture enabled the laboratory production of 'test-tube babies'. In turn, popular representations of tissue culture impacted on the ways that scientists discussed, used and even obtained tissues.

Attitudes to tissue culture were not polarized between 'science' or the 'public', then, but traversed and linked them. The scientists who used tissue culture were only one group in a dynamic network that also comprised journalists, authors, documentary makers, anti-vivisection and pro-life groups, bioethicists, lawyers and politicians. Each of the following chapters highlights the mutual interplay between these groups: detailing how tissue culture acquired significance at the intersection of scientific and social worlds, and catalysed new relations between them. By highlighting these historical interactions, I hope to provide much needed balance and perspective for current debates.

2

'Make Dry Bones Live': Tissue Culture at the Cambridge Research Hospital

This chapter shows how tissue culture became a high-profile tool in Britain during the 1920s and 1930s, and why it was particularly associated with the Cambridge Research Hospital. Although this was by no means the first or the only institution to use the method in this period, the medical and popular press identified it as the 'home of tissue culture experiments'.[1] It became famous, in part, because Hospital scientists, including its founder, Thomas Strangeways, aligned their tissue cultures with the changing concepts of time, the body and mortality that defined the modernist culture of the late nineteenth and early twentieth centuries.[2] Historians have detailed how these changes were driven by the scientific work of Albert Einstein and Marie Curie, by new movements in the arts and literature and by new visual technologies such as the cinema; but biologists also played a major role, by treating organisms and their parts as 'technologies of living substance'.[3] In a mindset that Philip Pauly identifies as 'recognizably modernist', they challenged the nature of mortality, by framing aging and death as manipulable and preventable.

Strangeways and his colleagues aligned themselves with these modernist ideas, publicly arguing that tissue culture redefined mortality and the body by allowing cells to exist independently of, and to even outlive, the organism from which they came. They argued that tissue culture made 'dry bones live', and sought to prove this by 'reviving' sausage meat *in vitro* and making cinematic films of cells growing outside the body. Medical and popular sources agreed that this work illustrated the 'indeterminate nature of death' and claimed, like Strangeways and colleagues, that tissue culture made cells and tissues immortal.

But tissue culture also contributed to the pervasive sense of anxiety and impending catastrophe that was characteristic of the interwar mindset.[4]

4

Popular reports outlined how it would leave readers 'fascinated – and a little frightened', and warned that it harboured a 'nightmare worthy of H.G. Wells'. This unease linked concerns surrounding the destructive capabilities of science to ongoing fears surrounding biological degeneration. Biologists and journalists argued that maintaining cells apart from the ordered environment of the body would foster their regression to a degenerate, atavistic state. These claims played out in newspaper and fictional accounts of rampaging tissues that were driven by a primordial urge to replicate and digest, which engulfed bystanders, cities and even whole societies.

Constructing tissue culture

From the outset, tissue culture embodied the experimental belief that 'biology' involved acting on, and not simply observing, natural phenomena. This mindset was promoted by mid nineteenth century figures such as the French physiologist Claude Bernard and the British scientist Thomas Henry Huxley, and it was later endorsed by the Cambridge physiologist Michael Foster and the German physiologist Jacques Loeb. It originated in physiology, but soon permeated bacteriology, embryology, pathology and zoology.[5] This experimental ethos treated organisms as raw materials that could be dismantled and reformed in novel ways; and during the late nineteenth and early twentieth centuries, experimental biologists began to re-assess the nature of living material, with a view to controlling natural phenomena. They isolated and disrupted the cells of developing embryos, grafted healthy or cancerous parts between animals and induced unfertilized eggs to divide by altering the chemical composition of their environment.[6]

In 1907, the Yale embryologist Ross Harrison drew upon this experimental tradition when he dissected out nerve cells from a young frog embryo and placed them in a sample of clotted lymph on a cover slip, which he then inverted over a depression slide and sealed with paraffin. He did so in order to settle a long-running dispute regarding the formation of nerve fibres.[7] Embryologists had long debated three competing mechanisms: that nerves were the product of a chain of cells, that they extended from a single cell, or that they grew between intra-cellular bridges. Advocates of all three positions had hitherto relied on microscopic analysis of histological specimens to support their particular stance. They would kill embryos at specific stages of growth, before sectioning, staining and analysing their tissues under a microscope. But the complicated network of cells scientists observed in these fixed specimens proved ambiguous.[8] Harrison sought clarification by attempting to isolate the nerve cell from its complex bodily environment, and his approach appeared successful. The 'striking' behaviour he claimed to observe *in vitro*, with fibres extending from the cell bodies, suggested that nerve fibres did indeed originate from a single cell.[9]

In a 1911 lecture, Harrison located his new technique in the experimental tradition, claiming it was 'the logical conclusion of the underlying idea of the method of elimination and transplantation of parts'.[10] But as Hannah Landecker notes, it marked an important divergence in one respect. Though biologists had maintained organs apart from the body before, they generally assumed that they would only survive briefly after their removal. Harrison, on the other hand, demonstrated that cells could be induced to thrive apart from the body – to grow and to divide as *in vivo*. His experiment opened the possibility that the body was no longer essential to the survival of its constituent parts.[11]

Harrison's new technique soon became associated with the French surgeon Alexis Carrel, who worked at the Rockefeller Institute for Medical Research, in New York. Carrel was also an advocate of experimental approaches in biology, and was already renowned for transplanting organs and limbs between different animals, with the aim of finding optimal conditions for wound healing.[12] But unlike Harrison, who was not interested in the new method as an end in itself, Carrel sought to refine and improve it, in order to find the conditions that encouraged optimal growth and regeneration of tissue.[13] With his assistant, Montrose Burrows, he cultured hundreds of diseased and healthy tissues from warm-blooded animals and hospital patients, varying the nutrient media and glass apparatus in order to prolong their survival *in vitro*. By naming this new method 'tissue culture', Carrel betrayed his emphasis on control and enhancement. 'Culture', of course, was an established term in bacteriology, where it denoted the laboratory growth of microorganisms, and it appeared the logical name for a method that essentially involved cultivating tissue and cells from animals in the same manner as bacteria or fungae.[14] But during the early twentieth century, in both biology and agriculture, 'culture' also referred to the intersection of nature and technology: describing a process that transformed and improved natural phenomena.[15] As a biologist who believed that nature could be controlled and improved through experiment, Carrel may well have had this latter definition in mind.

Carrel also differed from Harrison in the spectacular way he sought to drown out criticism. In 1910, the French histologist Justin Jolly had belittled tissue culture by arguing that excised tissue and cells could not live apart from the body. Jolly believed excised material invariably died *in vitro*, and that Carrel and Harrison mistook its brief survival and death throes for healthy growth and division.[16] In 1912, Carrel responded by claiming that he had cultured cells from a chick embryo heart for 85 days, thanks to careful maintenance and regular transferal to fresh medium.[17] This sample's health was apparent in its luxuriant growth, and by the fact that the heart cells continued to pulsate rhythmically after three months apart from the organism. Where Jolly claimed that the body was essential to the survival of tissue and cells, Carrel argued that the body's structural complexity actually imposed death upon them. He concluded that cultured cells were 'liberated' from the body's constraints, and there was nothing to prevent them living 'indefinitely' *in vitro*.[18] The death of any tissue *in vitro*, he claimed, should be ascribed to 'preventable occurrences' such as infection, the accumulation of catabolic substances and exhaustion of the culture medium.[19]

In order to publicize the work they funded, the directors of the Rockefeller Institute often pointed journalists to papers in their *Journal of Experimental Medicine*, which included Carrel's new research.[20] From the column inches dedicated to his work on 'permanent' life, it is fair to say their tactic succeeded. What became known as Carrel's 'old strain' featured in more press reports than any other research undertaken at the Rockefeller Institute, and became synonymous with tissue culture in the United States. Newspapers habitually recorded its 'birthday', including its coming-of-age in 1933, and even ran a series of premature obituaries when it was believed to have perished in 1940 (the culture was eventually discarded in 1946, two years after Carrel had died in occupied France).[21]

Carrel's presentation of the 'old strain' reflected broader modernist impulses. As Landecker has outlined, his writing on *in vitro* immortality was influenced by the theories of the French philosopher Henri Bergson, who played a major role in the reappraisal of time during the early twentieth century. Bergson stressed the heterogeneity of time, arguing that physiological time, or *dureé*, differed from the absolute time expressed by clocks.[22] Carrel argued that he had distinguished between absolute time and physiological *dureé* by culturing and seemingly immortalizing tissue. As he stated in a 1929 paper in *Science*, whereas absolute time was 'inexorable and irreversible', physiological lifespan could now be manipulated by tissue culture.[23] Commemorating the old strain's tenth 'birthday' in the *Journal of Experimental Medicine*, Carrel's colleague Albert Ebeling reiterated that it 'was no longer subjected to the influence of time'.[24] Press reports similarly dwelt on how tissue culture separated biological lifespan from absolute time. The *Fort Wayne Gazette* reported how 'the startling discoveries that science has made concerning tissues indicate that man has it almost within his grasp to keep right on living', while the *Raleigh Register* claimed that 'it may become possible to suspend human life and to start again'.[25]

These claims also appeared in British medical and popular reports on the 'old strain'. A 1920 editorial in the *Lancet* stated that 'we are now justified in looking upon the cells of animals as potentially immortal'.[26] Dwelling on the 'spectacular' implications of Carrel's work, it claimed that tissue culture 'affords grounds for the hope of immensely prolonging the life of man'.[27] In 1924, the *Daily Mirror* also used the 'old strain' to question if death was inevitable. 'If a single organ can be kept alive', it asked, 'why not a combination of organs?' The *Mirror* concluded that human life was 'far from fixed' and may be extended 'indefinitely'.[28]

But despite the attention his work received in the medical and popular press, we must not presume that Carrel dominated British coverage of tissue culture. In fact, reports on the 'old strain' were outnumbered by coverage of research undertaken at a small Cambridge laboratory, which lacked the Rockefeller Institute's facilities and financial resources. Indeed, in a 1933 report on tissue culture, *The Medical Press* claimed that Cambridge, not New York, was 'the home of tissue culture'.[29]

Tissue culture at the Cambridge Research Hospital

The institution in question was a modest, residential-looking building set amongst an acre of land in Worts Causeway, Cambridge. Known as the Cambridge Research Hospital, it had initially lodged in a small semi-converted house in the Hartington Grove district, where two bedrooms were converted into wards and the outside coal-shed served as a makeshift laboratory. The Hospital had been founded and bankrolled in 1905 by a group of local doctors, known as the Committee for the Study of Special Diseases. Keen to study and treat chronic diseases such as rheumatoid arthritis, the Committee justified the establishment of the Hospital by arguing in the *British Medical Journal* that:

> Much and fruitful research has been done in acute diseases, but there are many chronic and widespread maladies which have not yet received the attention they deserve ... Patients suffering from these diseases, owing to the slow progress and long continuance of their illness, are not suitable for reception into general hospitals,

which as a rule (and rightly so) admit by general preference cases of the more acute diseases. The mode of origin or progress of these affections is very imperfectly known, and little can therefore be done in the way of their rational treatment. The great need of scientific research on these chronic diseases has been strongly felt for many years.[30]

Following an appeal in the local press, the Hospital admitted 94 patients in its first two years.[31] During their stay, each patient was allotted to one of five available beds, subjected to experimental treatment and examined to determine the possible causes of their affliction, as well as any nervous changes associated with it.[32] In order to determine the morbid anatomy associated with rheumatic conditions, Hospital staff also obtained large quantities of pathological samples from London and the nearby Addenbrookes Hospital.

At this stage, the Committee for the Study of Special Diseases ran the Hospital in their spare time. The most involved Committee member was the University of Cambridge pathologist, Thomas Strangeways Pigg Strangeways, who had the original idea for the Research Hospital and served as its director since 1905. Strangeways had trained as a pathologist at St. Bartholomew's Hospital, London, during the 1890s. He moved to Cambridge in 1897, when his mentor, Alfredo Kanthack, was appointed chair of pathology. Like Kanthack, Strangeways saw pathology as more than a mere ancillary to medicine, and believed it had a future as an experimental laboratory-based discipline.[33] Although Kanthack died of cancer after a year in Cambridge, Strangeways stayed on and became Huddersfield lecturer in pathology under the new chair, German Sims-Woodhead.[34] Woodhead shifted emphasis back to the diagnostic and service aspects of pathology, which may have caused a dissatisfied Strangeways to set up the Research Hospital.[35] Nevertheless, Woodhead was supportive, providing money and allowing Strangeways to spend most of his time at the Research Hospital.[36] It had no formal ties to the University, though prominent Cambridge academics such as Woodhead and Sir Clifford Albutt, regius professor of physic, served on its board of Trustees.[37]

Running the Research Hospital certainly occupied Strangeways; when he was not examining patients or gathering pathological samples, he was constantly seeking financial assistance. He obtained money from local doctors and pathology students, and all Hospital staff worked for nothing, but it seems that Strangeways often struggled to meet the annual running costs – estimated at £500 in 1907.[38] In 1908, the *Lancet* reported that the Cambridge Research Hospital had been forced to close 'for want of funds'.[39] Although Strangeways attracted enough money to reopen it in 1909, the financial situation remained precarious. In 1910, the *British Medical Journal* suggested that its work, 'important as it is from a scientific as well as a humanitarian point of view, has about it no sensational element that appeals to the public'.[40] In a later report, the same journal claimed that by pursuing research into an 'obscure and inglorious' subject, the Hospital would continue to struggle financially. However debilitating rheumatic conditions were, it seemed they 'lacked the sentimental element that so powerfully helps hospitals for children and consumption'.[41] In a promotional booklet sent to doctors, Strange-ways admitted that rheumatoid arthritis aroused less sympathy 'than the more sensational dramas of disease in which battle is waged against death'.[42] Despite this, in 1912 he managed to raise enough money to plan and build a new Hospital building, providing space for a larger patient cohort and equipped with dedicated laboratory facilities (the con-verted coal-shed having proved unsurprisingly inadequate).[43] This time, the Committee appealed for money beyond medical circles, raising the necessary funds largely thanks to private bequests and a donation of £700 from the philanthropist Otto Beit.

Following its opening in May 1912, work at the new Hospital pro-gressed as before: patients were admitted and observed, while Strange-ways collected tissue samples to analyse in his new laboratories.[44] The only notable difference was the appointment of Victor Norfield as a dedicated technician and general assistant.[45] However, this routine changed drastically after World War I, during which the Hospital was used to treat injured soldiers. From 1919, Strangeways became increas-ingly preoccupied with tissue culture, which he probably learnt from the technician he employed during the War, whilst Norfield was serving in France. This technician, a Mr. Sheldrick, had previously worked at Carrel's Rockefeller department, where he investigated the effect that various growth hormones had on chick embryo tissues.[46]

Strangeways initially employed tissue culture in order to better under-stand the cellular origins of rheumatoid arthritis, by culturing and observ-ing the growth of cartilage from adult chickens and chick embryos.[47] Yet he was soon using the technique to expand his research. In 1922, Strangeways published a detailed account of the growth and cell division of chick embryo cells *in vitro*.[48] Like Carrel, he also became interested in the method as an end in itself and published a tissue culture manual the following year, offering guidance on installing tissue culture apparatus, on sterile technique and on which material grew best *in vitro*.[49] As he became more interested in tissue culture, Strangeways stopped admitting patients to the Research Hospital and advised them to seek treatment at St. Bartholomew's instead. In 1923, he closed all observation wards and converted them into tissue culture laboratories.[50]

Later in the 1920s, promotional literature for the Research Hospital stated that Strangeways adopted tissue culture because it placed a potent 'new weapon in the hands of the biologist'.[51] Yet this conversion was not quite so straightforward; new funding was involved and we cannot appreciate Strangeways's shift to tissue culture without also appreciating the financial climate biologists faced immediately after World War I. In 1919, the Medical Research Council (hereafter MRC) was formed by a reconstitution of the 1911 Medical Research Committee.[52] The MRC's remit was to free biological science from clinical concerns and encourage research into fundamental scientific problems. Its inaugural secretary, Walter Morley Fletcher, was a graduate of Michael Foster's department of physiology at the University of Cambridge, where students were encouraged to adopt new experimental approaches.[53] Like Harrison and Carrel, Fletcher firmly believed that scientific and medical progress could only be attained through experiment, and he envisaged Cambridge as the centre of a growing research culture in British biology.[54] Soon after its formation, the MRC endowed the Research Hospital with a block grant and became its primary source of income. But unlike previous benefactors, it attempted to direct the Hospital's research activity: quickly pressing Strangeways to convert to tissue culture and abandon his work with patients.[55] As Strangeways conceded in a letter to Malcolm Donaldson, a London clinician and Hospital trustee, he had become dependent on MRC funds and realized that the only way to keep the Research Hospital open was to meet the Council's demands.[56] It seems, then, that his shift to *in vitro* work was borne more out of financial necessity than any faith in this 'new weapon'.

The MRC's faith in tissue culture no doubt stemmed from its visible association with the Rockefeller Institute, whose commitment to laboratory research was a marked influence on Fletcher.[57] It may also have stemmed from clinically promising work that Strangeways had already undertaken, with MRC funds, on the *in vitro* effects of radiation. Working with the physicist H.E.H. Oakley in 1921, Strangeways exposed cultured chick embryo cells to regular doses of radiation, and noted that damage was most pronounced whilst they were undergoing mitosis.[58] Strangeways and Oakley recorded that cells showed distinct cytological abnormalities and began to 'break down' after prolonged exposure to radiation. The following year, Strangeways extended this work in a collaboration with staff at St. Bartholomew's who were interested in radiological research. He and Norfield prepared cultures of chick tissue at the Research Hospital and drove them to London, where they were irradiated and examined by F.L. Hopwood, a physicist, Ronald G. Canti, a bacteriologist and pathologist, and Malcolm Donaldson, the gynaecologist and Hospital trustee. This so-called 'Strangeways team' was supplemented in 1925 when the radiobiologist Frederick Gordon Spear joined the Research Hospital.

At the same time, Strangeways extended his initial observations on cell growth, by investigating whether explanted limb buds and embryonic eyes grew as normal *in vitro*. This was prompted by the 1923 appointment of the young zoologist Honor Fell, and marked the beginning of a long tradition of developmental studies at the Research Hospital (which is discussed more in the next chapter).[59] In 1924 Strangeways even tried to culture whole chick embryos. Whilst Fell later recalled that he met with 'considerable success' in these experiments, they were not pursued in earnest until Conrad Waddington joined the Research Hospital in the 1930s.[60]

Figure 2.1   Staff and visitors at Cambridge Research Hospital, 1924. Back row (l–r): J.A. Andrews, H.B. Fell, V.C. Norfield, J.G.H. Frew. Front row (l–r): F.G. Spear, T.S.P. Strangeways, R. Chambers, R.G. Canti. Courtesy of the Wellcome Trust Library for the History of Medicine.[61]

But Strangeways's conversion to tissue culture coincided with growing criticism of the method. In 1923, the *Lancet* claimed that whilst
it had 'promised much, it must be confessed that its fruits have
hitherto been meagre and far from encouraging'. It complained
that researchers often utilized tissue culture for vague reasons and
gave 'no hint of the direction in which these investigations may
lead'.[62] The following year, another dismissive *Lancet* piece remarked
that 'we do not yet know whether we have in the method a
means of solving certain fundamental biological questions ... or
a barren though brilliant laboratory technique'.[63] This scepticism
was clearly ingrained. Opening a 1924 speech to the British
Medical Association (hereafter BMA) in Bradford, the normally confident Alexis Carrel was forced to admit that attempts to apply
tissue culture to medicine 'did not generally meet with great
success'.[64]

Strangeways and his growing band of colleagues were by no means
the first researchers to have utilized tissue culture in Britain. In 1914
the brothers John and David Thomson, from the Royal Society's Marcus
Beck laboratory in London, published papers on the cultivation of
chick embryo and human tissues.[65] The same year, the bacteriologist
Albert J. Walton, who worked at the London Hospital, sought to determine the effects of various nutrient media on rabbit tissue *in vitro*.[66]
Both Walton and David Thomson had, notably, travelled to New York
to receive practical training from Alexis Carrel. And in the 1920s
A.H. Drew, from University College London, and Edward Willmer,
from the University of Manchester, published advice on tissue culture and detailed the behaviour of *in vitro* material.[67] Strikingly,
however, none of these figures appeared in popular coverage of tissue
culture during the interwar period. When they described the emergence of tissue culture in Britain, newspapers and popular books
invariably turned to the Cambridge Research Hospital, not London or
Manchester.

Strangeways and his Research Hospital colleagues appeared regularly in popular coverage because they went to great, and often sensational, lengths to overcome the growing criticism of tissue culture.
Whilst Willmer simply argued that scientists would 'reap considerable
benefits' by persevering with tissue culture, researchers at the Research
Hospital went much further and argued that it permitted new ways of
observing, controlling and even reconfiguring biological phenomena.[68]
Crucially, they did so in newspapers and on the radio, not just in
medical lectures and published papers.

Part of their support for tissue culture involved establishing a sense of superiority over the pathological samples that were still more commonly used by scientists. Like Carrel, who demeaned anatomical specimens as 'only an artefact ... nothing but useful abstractions', Frederick Spear claimed in a 1928 lecture that using morbid tissues after working with tissue cultures was like 'passing from the village green gay with the liveliness of children at play into the gloom and morbidity of the borough mortuary'.[69] Spear argued that in contrast to fixed samples, which captured a tissue or physiological process *in* time, tissue culture granted scientists the ability to observe 'the thing itself' as it unfolded *over* time.[70] In distancing tissue culture from pathological samples, he also emphasized how it effaced death. Put simply, tissue culture was a superior and more objective technique because it made 'dry bones live'.[71]

Strangeways permeated his books, articles and lectures with a similar sense of vitality. In a 1922 paper presented to the Royal Society, he claimed that tissue culture demonstrated how the cell and its internal components were 'never at rest'. Later in the talk, Strangeways compared pigment rods in eye cells to 'guinea pigs in a run', and the 'writhing movement' of chromosomes during cell division to 'eels in a box'.[72] In his 1924 book on *Tissue Culture in Relation to Growth and Differentiation*, he stated that tissue culture had opened up 'a new field of investigation' by demonstrating how the living cell was 'essentially dynamic'.[73] During a 1926 lecture to the BMA, he argued that anyone viewing a cultured cell would be instantly struck by its 'individuality and independent behaviour, and would realize, as perhaps never before, that the cell, properly speaking, was an organism complete in itself'.[74]

This belief that a cell was a potentially independent organism had a long history, and was one of several standpoints in a debate on the status of cells. Following the formulation of the cell theory in the mid nineteenth century, some biologists claimed that organisms were nothing more than organized colonies of cells. More holistically minded biologists opposed this claim, however, and argued that structure and function could only be understood in terms of a complete whole. As Strangeways's lectures indicate, the development of tissue culture supported the view that bodies were essentially congeries of potentially autonomous cells.[75] And if each cell was an autonomous organism, then to Strangeways 'the death of the body in no way necessitated the death of its component cells'.[76]

By 1926, the possibility of experimentally manipulating lifespan had come to embody both the power and promise of experimental biology. It was not only evident in Carrel's 'old strain', but was also evoked by the zoologist Julian Huxley's experimental maturation of the axolotl and by Steinach and Voronoff's apparent rejuvenation of elderly humans through gland grafting. By engaging with these broader efforts, which Huxley called a 'search for the elixir of life', Strangeways brought his Research Hospital to public attention in the late 1920s.[77] During his BMA address and a Cambridge University lecture on 'Death and Immortality', he claimed tissue culture proved that all cells were autonomous and potentially immortal.[78]

This was by no means a radical claim. In a 1920 report on the 'old strain', the *Lancet* labelled it a 'commonplace of thought', and in 1922 Julian Huxley claimed that most biologists believed 'only the system as a whole is doomed to death'.[79] There was also nothing new about Strangeways's claim that tissue culture allowed cells to be 'kept alive in active reproduction for a period considerably longer than the animal's entire span of life' – except that he did not refer to the 'old strain' once.[80] Instead, Strangeways demonstrated the potential immortality of cells by employing his own practical demonstration, which newspapers held as conclusive proof of the indeterminate nature of death, and of his status as 'the master of investigation in living cells'.[81]

In his BMA and Cambridge lectures, Strangeways recalled the experience of an unnamed colleague at the Research Hospital who required kidney tissue. He outlined how thanks to the 'slackness' of the dustman, a dead experimental rabbit had lain for two days in a laboratory bin. Strangeways stated how his colleague then removed the dead rabbit, dissected its kidney and washed tissue samples in iodine. He claimed that once explanted *in vitro*, these tissues 'grew perfectly'.[82] For Strangeways, this proved that cells continued to live for days after the seeming 'death' of the body.

In his Cambridge lecture on 'Death and Immortality', Strangeways went a step further to demonstrate how cells survived after the body had seemingly died. He urged his students to imagine passing an apparently dead body through a mincer, before fashioning it into a string of sausages. 'The body', he continued

transformed into sausages does not become dead for days or, if kept in cold storage, weeks (unless life is abruptly terminated in the frying pan). Assuming the procedure has been carried out under

aseptic conditions and that no toxic seasonings have been added it
will be found that on cultivating pieces of the mince meat *in vitro*,
active division and migration of the cells will take place.[83]

Strangeways then revealed a culture he had made from minced sausage
meat purchased at the local butchers, in which he claimed the cells
moved and continued to proliferate.

Like the 'old strain', we may designate this sausage culture as an
inherently modernist entity. It offers a clear example of how biologists
forced biological matter to do and be something novel, resonating with
shifting understandings of the body and lifespan. But the immortal
sausage offered an even more spectacular engagement with these concerns than the 'old strain'. For while Carrel harnessed the developmental potential of embryonic cells – long associated with vitality,
regeneration and laboratory science – Strangeways forced his audience
to reassess the nature of an everyday object, which few would have
regarded as alive in any sense. While Carrel demonstrated the vitality
of cells at the start of life, Strangeways did so at its end. He concluded
that the only factor that may curtail the life of these sausage cells was
the 'patience and longevity of the researcher' who cultured them.[84]
This claim was unfortunate in light of events later that day. After cycling
home from his lecture on 'Death and Immortality', Strangeways suffered a brain haemorrhage and was found unconscious in an armchair.
He died on the 23 December 1926.

'The malignant entity': Tissue culture in the popular sphere

In contrast to the 'obscure and inglorious' work performed there before
the 1920s, Strangeways's research on tissue culture certainly raised the
profile of his Research Hospital. Ironically, this began in his obituaries.
The *British Medical Journal* claimed that Strangeways's sausage culture
perfectly illustrated the 'indeterminate character of death', and similar
claims circulated in press reports.[85] A 1927 piece syndicated for US newspapers described the sausage culture as a 'spectacular instance of how
life may persist in something which is apparently dead'.[86] Newspapers
claimed that Strangeways had transformed notions of lifespan, by 'res-
cuing' cells from the both the sausage and the death of the original
animal.[87] Some papers that ran this story even complemented it with a
visual re-imagining of the experiment (see Figure 2.2).

But while newspapers brought Strangeways posthumous attention,
his Research Hospital faced an uncertain future. At an emergency

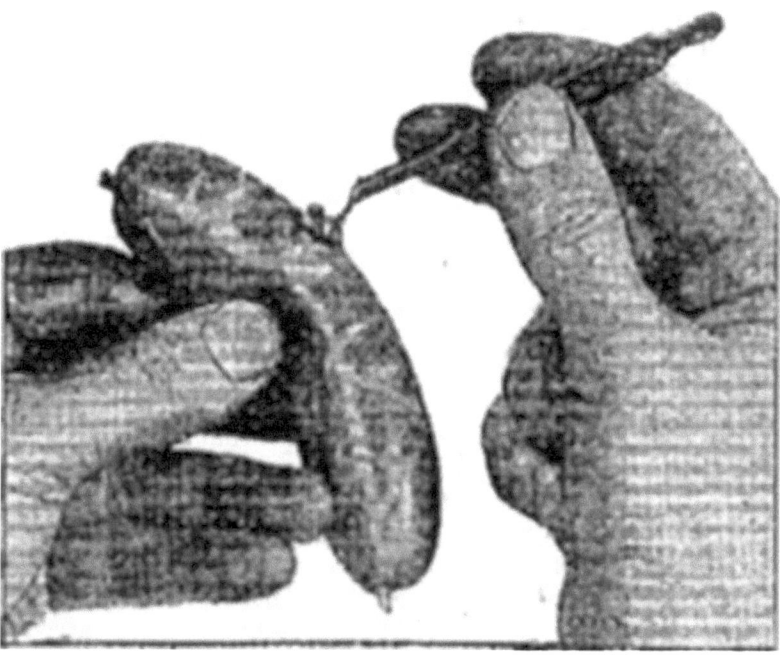

Figure 2.2   Illustration from an American newspaper the *San Antonio Light*, with Strangeways demonstrating that 'even a sausage contains living cells'.[88]

meeting of its three remaining trustees in January 1927, only one lobbied for its survival, while two pressed for closure and the incorporation of researchers into Cambridge University.[89] They eventually allowed the Hospital's survival, provided it bolstered its income by providing courses in tissue culture techniques.[90] In 1929, the trustees appointed the 28 year-old Honor Fell as director of a newly renamed Strangeways Research Laboratory, which immediately began to run a practical course on 'Tissue Culture and its Applications'.[91]

The early years of Fell's directorship were marked by expansion in several areas. The number of scientists at the Strangeways Research Laboratory increased, as researchers like Conrad Waddington and Alfred Glücksmann were allocated laboratory space and taught tissue culture.[92] The MRC income was supplemented by grants from the British Empire Cancer Campaign and the Rockefeller Foundation, which paid for new researchers, new equipment and the construction of more laboratories.[93] And the late 1920s and early 1930s also saw a marked growth in popular coverage of tissue culture.

This latter increase was partly due to Honor Fell's enthusiasm for what she described, in correspondence with Malcolm Donaldson, as 'getting ideas into the public's skull'.[94] In contrast to Strangeways, who was shy on account of poor hearing, and confined his promotion of tissue culture to published material and medical lectures, Fell openly endorsed the method in a variety of public forums. She regularly lectured to schoolchildren and the public, and in January 1930, she gave a BBC radio talk on 'The Life of a Cell'. Fell's popular activities formed part of an increase in the amount of 'public science' during the interwar years, though her output paled in comparison to individuals like Julian Huxley, the geneticist J.B.S. Haldane and the embryologist Joseph Needham. The increase in 'public science' during this period resulted from a number of ideological and professional factors: including leftist political commitments, support for a greater scientific role in national planning and a desire to restore public faith in science, which had been shaken following the deployment of poison gas and tanks in World War I.[95] There were certainly clear professional motives behind Fell's promotion of tissue culture. As she told Donaldson in 1934, she regarded public communication as useful 'propaganda' that would inculcate optimism in tissue culture and secure a financial future for the Strangeways laboratory.[96]

Fell's BBC talk was certainly good 'propaganda' for tissue culture, stressing its relevance to science and medicine. Fell claimed tissue culture was an important technique because it separated cells from the 'confusing influence' of the body. Before tissue culture, she argued, it was difficult to ascertain whether the changes observed following, say, irradiation were due 'to some general reaction of the body, or to the direct effect of the rays on the cells, or both'.[97] Fell stated that tissue culture now allowed scientists to detect the effect of particular reagents with 'remarkable precision'. This was reiterated days later in a *Sunday Express* report written by the physicist Sir Oliver Lodge, who praised the way that Fell and colleagues observed biological phenomena 'without the complications inevitable when the tissues form part of a living animal'.[98] Like Fell, Lodge demanded financial support for tissue culture, arguing that: 'It seems difficult to exaggerate the importance of such a discovery ... funds must be forthcoming, the work must grow and extend, and more laboratories must be established'.[99]

But this stress on medical utility did not mean that tissue culture's more sensational implications were toned down. Lodge also claimed that it reconfigured lifespan by facilitating the immortality and limitless growth of *in vitro* cells. Detailing Strangeways's culture of sausage mince, he claimed that before tissue culture 'no-one could have supposed that ... a fragment of tissue removed from the dead body of an animal could be revived and preserved in a living condition for an indefinite period'.[100] An identical claim appeared in *The Science of Life*, which was written by H.G. Wells, his son George Phillip Wells and Julian Huxley, and was serialized throughout 1930. Huxley and the two Wells hailed tissue culture as one of the new biological procedures that were 'bringing life under control'.[101] To illustrate this power, they detailed how it allowed cells to live 'for periods of time exceeding the individual life-span of the species from which they were taken'.[102] To substantiate this claim they notably drew on the work of Strangeways, not Carrel. If Strangeways had been living in the time of Julius Caesar, they noted, 'fragments of that eminent personage might, for all we know to the contrary, be living now'.[103] Honor Fell also confronted her BBC listeners with this changing view of the body and lifespan. She encouraged them to see themselves as 'a colony of cells', and appreciate how these constituent parts could be removed and made to live independently:

I dare say you think that if a piece of your flesh were cut off by a surgeon it would be dead as soon as it parted from your body. But, as a matter of fact, this is not the case. Of recent years science has shown that not only does animal and human flesh – or tissue, to use its scientific name – remain alive for quite a long time after death, but that in some cases it may actually be made to go on living and growing apart from the body for months and even years.[104]

Tissue culture was also brought to public attention thanks to the cinematographic films that Ronald Canti, part of the 'Strangeways team', had been making since 1926. These films originated in Strangeways's desire to capture the movement of cells in tissue cultures, overcoming his reliance on static photos to convey the 'writhing movement' he claimed to see *in vitro*. Shortly before Strangeways died, Canti built a cinematograph at his home in Hampstead, London. This consisted of a cine-camera positioned above a microscope, which Canti positioned in front of a large source of illumination and surrounded with a tube of velvet to cut out extraneous light. He then encased the whole apparatus in an incubator to maintain a constant temperature, and dampened the vibrations from a nearby Underground line by resting it on alternate layers of concrete and rubber. Canti finally placed a shutter between the light and the microscope, with a timing mechanism arranged to capture images at intervals of 20, 30 or 60 seconds and immediately wind on the cine-camera film. To overcome the slow rate at which events proceeded *in vitro*, and to convey a sense of rapid change, the captured cine-film was projected at a rate of 16 images per second. Strangeways established tissue cultures at the Research Hospital and transported them to London, where Canti placed them in his incubator and began to film them *in vitro* (after Strangeways died, Victor Norfield delivered the cultures).

These so-called 'time lapse' films of tissue culture had clear modernist resonance: the projection manipulated time and showed a technique that was believed to reconfigure the body and mortality.[105] Unsurprisingly, then, Canti's films received considerable popular attention. After their first screening to the Royal Society in 1927, *The Times* claimed that images of cells growing and dividing apart from the body had a 'striking psychological effect' on spectators and left them 'gasping with astonishment'.[106] Canti's film raised questions of 'great interest' regarding the nature of time and lifespan – or what *The Times* referred to, with Bergsonian overtones, as the 'fallacy' of normal time. For the *Manchester Guardian*, meanwhile, the films were simply a 'revelation'.[107] Other reports detailed the intricacies of Canti's apparatus, and outlined how his films allowed clinicians and scientists to directly study the effect radium had on cells. Like Fell, Canti also gave a BBC radio talk that used his films as propaganda for the tissue culture and the Strangeways laboratory. Here, he argued that tissue culture was 'a great advance' that allowed scientists to 'study the normal history of the cell, in its living state'.[108] Canti stressed that cinema films of tissue culture illustrated these benefits 'to a great extent'. He also took pains to remind listeners that the films were projected at a quicker rate than normal, and that 'the cells of your own body do not in fact move with the great speed which is suggested on the screen'.[109]

Popular coverage peaked in 1932, when Canti introduced his films to the Prime Minister Ramsay MacDonald and the Duke of York at 10 Downing Street.[110] This illustrious audience watched a new film, entitled 'The Cultivation of Living Tissue', that Canti had made for both scientific and popular audiences by editing together existing recordings to construct a narrative of 'life' *in vitro*.[111] The first frames showed normal chick cells growing and dividing; this then cut to the organized growth of rabbit and chick organs; and the film concluded with a shot of cancer cells shrinking and dying after treatment with radium.[112] Canti superimposed a clock onto the corner of the screen to illustrate how much time actually elapsed in each scene. After he died from pneumonia in 1936, a commemorative portrait was added as the final frame. The deliberately fabricated nature of 'The Cultivation of Living Tissue', and the sizeable news coverage it attracted, demonstrates how scientific films in this period could simultaneously exist as experiments, professional propaganda or pure entertainment.[113] And it also illustrates how tissue culture was constructed and given meaning across professional and popular domains.

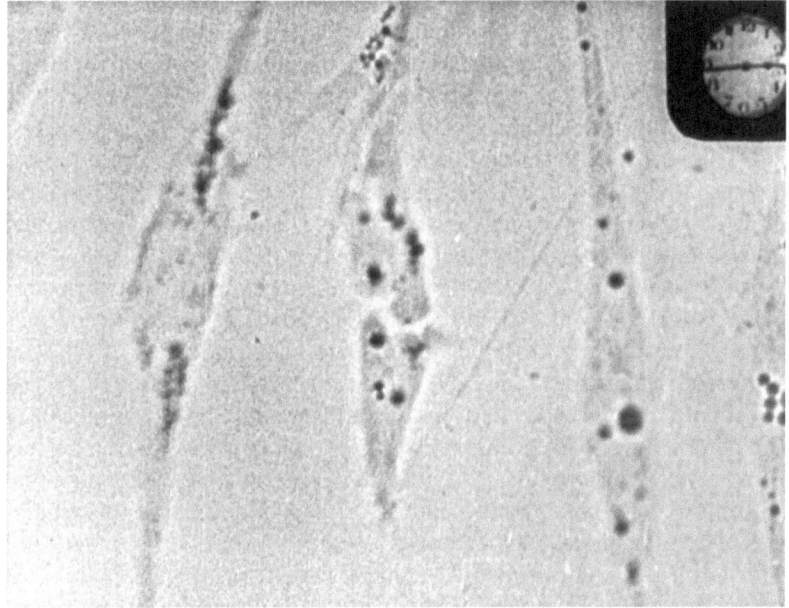

Figure 2.3   Still from 'The Cultivation of Living Tissue' (1932), showing chick fibroblasts growing *in vitro* and the clock in the right-hand corner. Reproduced courtesy of the British Film Institute.

But whilst the activities of Fell and Canti certainly increased the public profile of tissue culture, we must not presume that popular sources simply reiterated their support for the method. Instead, newspaper reporters and fictional writers often appropriated scientific rhetoric to frame tissue culture as a threatening and sinister technique – symbolic of the danger that many felt lurked within modern science and technology. A 1937 *Paris Daily Mail* report on the Strangeways laboratory gave a sense of this unease, warning readers that they would be 'fascinated – and perhaps a little frightened' by tissue culture.[114]

This ambivalence placed tissue culture within the growing inter-war criticism of scientific practices and technologies.[115] The perceived novelty of much experimental science, as well as other modernist inno-vations like the cinema, often generated a mixture of fascination and fear among interwar commentators. Popular and scholarly appraisals oscil-lated between admiration and insecurity, with commentators warning of harmful consequences for the general population and lamenting the loss of a more innocent, often rural, past. Though disquiet with science and technology predated the interwar period, it increased after the deadly application of poison gas and tanks during World War I, which weak-ened the Victorian equation between science and social progress. While writers like H.G. Wells explored the negative connotations of science during the late nineteenth century, a greater number of interwar writers, radio dramas and films now framed it as a potent threat to civilization. Over half of the horror films released between 1931 and 1936, for instance, presented science as the route to catastrophe.[116] The author Aldous Huxley, Julian Huxley's younger brother, encapsulated this unease in 1933 when he questioned whether scientific progress inevitably entailed destruction, and whether 'the very arts and sciences which we have used to conquer nature have turned on their creators and are now conquering us'.[117]

In order to portray tissue culture as a dangerous technique, news-paper and fictional accounts appropriated two scientific claims regarding the behaviour of tissues and cells *in vitro*. The first was that tissue cul-ture conferred immorality on cells, by effacing what Fell called the 'enormous complexity' of the body.[118] The second was that this 'libera-tion' encouraged cells to regress to a primitive state in tissue culture. This latter claim drew on the scientific belief that the structural and organizational complexity of the body was a mark of evolutionary progress. E.N. Willmer, for one, claimed that evolution could be charted by studying the increasing regulation in multicellular bodies, where the autonomy of constituent cells was subordinated for 'the good of the whole'.[119] In his 1923 *Essays of a Biologist*, Julian Huxley similarly extolled the 'delicate mechanism for co-ordination' found in animal bodies.[120] And in his radio talk, Canti argued that this 'communal' behaviour was 'responsible for the balance and control of our bodily functions, and indeed for our very existence'.[121]

Scientists believed this 'subordination' controlled the primitive urges that Darwinian and Freudian theory suggested lurked within the body and mind.[122] Any breakdown in this control was held to be a mark of evolutionary degeneration – of recapitulation to a primitive and dangerous state. Scientists held the free movement and unregulated division of cells to be degenerate behaviour, characteristic of primitive unicellular organisms like protista and amoeba. They also cited it as a prime cause of cancer, where hitherto subordinate cells regressed to a degenerate malignancy. To Julian Huxley, the malignant changes associated with cancer were marked by 'emancipation from the controlling bond which regulates the growth and harmony of the parts in the whole, and a career of unlimited, unregulated, growth and reproduction'.[123]

Unsurprisingly, then, scientists questioned whether removing tissues and cells from the body facilitated their regression to an atavistic state. Huxley argued that tissue culture demonstrated just how easily 'the cells which have been specialised to perform all the diverse functions of [the] body can revert to a primitive, embryonic state'.[124] And Willmer was clear that scientists encouraged degeneration by culturing cells *in vitro*. Like Huxley, he noted how cultured cells began to grow in an 'uncontrolled' manner and lost the morphology they had whilst in the body – a phenomenon which scientists labelled 'de-differentiation'.[125] Thomas Strangeways also linked de-differentiation to evolutionary degeneration, noting how the unregulated movement of cells in tissue culture resembled the 'primitive' behaviour of unicellular organisms and bacteria.[126] The Oxford histologist H.M. Carleton, meanwhile, likened the behaviour of cultured cells to malignant tumours.[127]

These observations were not confined to scientific sources. In a 1926 report for the *Manchester Guardian*, Julian Huxley outlined how cultured cells recapitulated to more 'embryonic' form.[128] And in *The Science of Life*, he and the two Wells described how Strangeways encouraged cells to 'desert' the body's regulation, fostering their reversion to a 'primitive' state that was characterized by a loss of function, free movement and unlimited growth.[129] In contrast to the 'disciplined' behaviour of cells in the body, they noted how cultured cells 'wander about where they like'. While a well-regulated organ like the brain or liver was thus 'like the City during working hours', they claimed a tissue culture resembled 'Regent's Park on a bank holiday, a spectacle of rather futile freedom'.[130]

Scientists like Strangeways and Willmer framed this 'degeneration' as a technical problem that prevented analysis of cells in their functional *in vivo* state. Several popular sources, on the other hand, used it to frame tissue culture as a direct threat to people and society. An *Observer* article on Canti's films, for example, compared the 'unrestrained growth' of cells onscreen to a 'tumour that went on and on at the expense of the individual in which it formed'.[131] Other accounts took this parallel further. In 1926, the science fiction magazine *Amazing Stories* published a short story by the popular writer Otis Adelbert Kline, entitled 'The Malignant Entity' (although *Amazing Stories* was published in the United States, it sold well in Britain and contained stories by British writers like H.G. Wells). The 'Malignant Entity' was a rapidly growing tissue culture that devoured several hapless individuals, including the scientist who cultivated it, thanks to its 'primitive desire for food and growth'.[132] These degenerate impulses were only curtailed after a poisonous chemical was poured onto its nucleus (see Figure 2.4).

Other narratives imagined scenarios where cultured tissue wreaked even greater havoc. In a 1932 report on Canti's films, the weekly tabloid *Tit-Bits* argued that the 'staggering' proliferation of cultured cells could result in 'the earth overwhelmed by a mass of protoplasm'.[133] This possibility was conveyed by an image of rampant tissue destroying a cityscape and scattering terrified people in its wake (see Figure 2.5). Such an outcome, the report continued, was a 'nightmare worthy of the imagination of H.G. Wells'. In 1937, the American radio play *Lights Out* foretold the destruction of civilization in the same way, with cities and their inhabitants engulfed, tellingly, by a culture of chicken heart that escaped a New York laboratory. The play closed with a band of survivors circling this growing mass in a plane, with their fuel about to run out, and one character proclaiming that the 'end has come for humanity: not in the love of atomic fusion; not in the cold winds of interstellar combustion; not in the winds of white cold pilings. But in cells. That creeping, grasping flesh below us'.[134]

Stories of rampaging tissue cultures embodied the interwar belief that potent dangers lurked beneath the veneer of scientific 'progress'. It is clear that they were not the creation of what contemporaries damned as 'feather-brained' tabloids or science fiction magazines.[135] Rather, they drew on scientific claims that tissue culture effaced the body and mortality, which had material support in the form of the 'old strain' and the culture of sausage mince. Yet whilst Carrel and Strangeways framed the effacement of mortality as evidence of tissue culture's potential, some journalists and popular writers framed it as a threat to individuals and society. Indeed, unendingly spreading, immortal tissues came to serve as a metaphor for the dangerous potential of biology in general. Concluding a 1938 report on the possibility of synthetic blood, hormonal rejuvenation, organ transfer and the *in vitro* production of 'chemical babies', *Tit-Bits* sounded a familiar note of caution when it urged its readers to 'remember that chicken heart which went on growing and growing...'[136]

Figure 2.4   Illustration to 'The Malignant Entity' in *Amazing Stories*. The degenerate cell is shown being killed by the addition of poison to its nucleus. Note the corpses below this tissue culture.

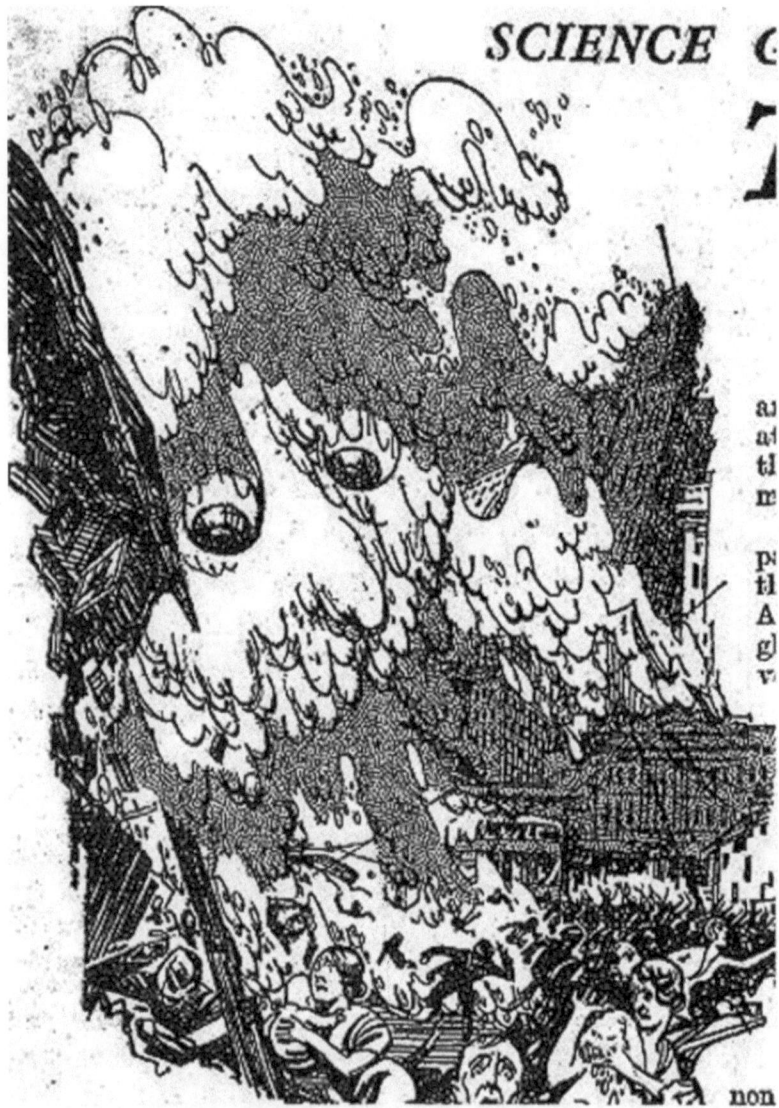

Figure 2.5   'The earth overwhelmed by a mass of protoplasm'. *Tit-Bits* illustrates the danger of tissue culture. Courtesy of the Wellcome Library for the History of Medicine.

By reflecting fascination with the reconfiguration of mortality and the body, and highlighting anxiety surrounding advances in science and technology, these debates firmly embedded tissue culture within the modernist climate of interwar Britain.[137] Moreover, tissue culture was a crucial node in the *making* of this broader climate, since it regularly served as a vehicle through which scientists, journalists and popular writers discussed the reconfiguration of the body and mortality – for better or worse. There is also good evidence that popular representations of tissue culture impacted on scientists. From the mid 1930s, biologists began to protest against overly sensational portrayals of tissue culture and urged their colleagues to present it in a more measured light. In a 1935 textbook, Edward Wilmer stated that tissue culture should never be 'regarded as anything more than a method' that had simply 'furthered knowledge of the processes involved in the normal development of the organism'.[138] He argued this balanced view was needed to counter the fact that 'tissue culture has in the past been the object of much misrepresentation and of many false conceptions'.[139] Notably, Wilmer conceded that these misrepresentations often arose from scientists, who had responded to criticism that tissue culture 'contributed nothing to science' by portraying it as 'a magic key to the understanding of life processes'.[140]

Conclusions

This chapter has shown how early discussion of tissue culture reflected, and helped instantiate, the modernist landscape of interwar Britain. Though several other researchers and institutions employed tissue culture during this period, staff at the Cambridge Research Hospital became publicly associated with the method by regularly engaging with the contemporary reappraisal of time and living matter. In order to promote the 'powerful new weapon' that became the mainstay of their research, and the best hope of their institution's survival, they argued that it conferred independence and immorality on cultured tissue and cells.[141] Their research and spectacular claims appeared regularly in popular sources; and these popular representations, in turn, impacted on the scientific presentation of tissue culture. This, then, is the interplay between economic considerations, material practices, modernist ideas, human actors and non-human objects that permits us to talk about the *co-production* of British science and culture.[142]

There was no dichotomous gap between scientific and popular portrayals of tissue culture in this period. Positive and negative representations of the technique were both structured by common concerns, sharing the belief that it conferred immortality and limitless growth upon cells *in vitro*. Any rigid distinction between the 'scientific' and 'popular' is undermined by the fact that scientists often promoted tissue culture in popular outlets, and that their arguments underpinned sensational popular accounts. The editorial policy of periodicals such as *Tit-Bits* and *Amazing Stories* was not to fabricate wild claims, but involved repackaging scientific ideas for mass consumption.[143] And scientific claims regarding autonomous and immortal cells clearly made good copy, chiming with modernist concerns and blurring the boundary between scientific fact and fiction.

3
'Could You *Love* a Chemical Baby?' Organ Culture in Interwar Britain

This chapter looks at popular coverage of another aspect of tissue culture research during the interwar period, focussing on the cultivation of whole organs and embryos *in vitro*. Although this line of research was marginal compared to the growth of 'de-differentiated' cells, it was scientifically important and culturally resonant. The ability to grow embryos and their parts outside of the egg or the mammalian uterus helped answer many questions surrounding growth and differentiation, and allowed scientists to determine the effects various of chemicals and hormones on development. From the 1940s, the method which scientists labelled 'organ culture' became an important approach in embryology, cell biology, biochemistry, endocrinology, toxicology and physiology.[1]

This chapter again centres on the Strangeways Research Laboratory – partly because little organ culture was done elsewhere during the interwar years, either in Britain or further afield. Indeed, when greater numbers of British and European laboratories began to adopt the method in the 1940s, the work was usually conducted by scientists who had worked or trained at Cambridge.[2] But the Strangeways Research Laboratory is not just relevant as historical centre for organ culture techniques. It again highlights how tissue culture was a visible and contested technique. Several popular writers and journalists drew on work performed there to claim that organ culture allowed the production of so-called 'test-tube babies': linking the method to contemporary debates over new mass-production methods, eugenics, birth control and the changing social status of women. The prospect of cultivating children in the laboratory attracted the same mixture of admiration and ambivalence as other modernist ideas and innovations. Proponents of eugenics and birth control argued that 'test-tube babies' would counter declining population health, and free women from the rigours of pregnancy and childbirth. Others, however, portrayed test-tube babies as symbols of the negative consequences of eugenics, feminism and industrial mass-production techniques. Opponents of feminism and women's suffrage, for example, claimed the production of babies *in vitro* would render men biologically and socially 'superfluous'.[3] And authors like Aldous Huxley framed test-tube babies as symbols of human subordination to industrial culture, envisaging scenarios where couples were displaced from reproduction, and dehumanized babies were turned out on production lines like Henry Ford's Model-T car.

Like the 'old strain' and the immortal sausage, 'test-tube babies' highlight the interaction between popular and professional portrayals of tissue culture, as well as the porous boundary between science 'fact' and 'fiction'. Numerous popular accounts cited the development of organ culture to legitimate their predictions of test-tube babies. But many of these were written by scientists; and their impact was compounded when figures like Thomas Strangeways publicly linked organ culture to test-tube babies. Indeed, Strangeways's 1926 claim that organ culture supported 'the test-tube baby' appears to be the first public use of the term. As popular accounts and scientific endorsement of test-tube babies increased, newspapers claimed that Honor Fell and her Strangeways colleagues were taking 'steps to the *Brave New World* visualized by Aldous Huxley'.[4] These debates also demonstrate how popular representations can, in turn, impact upon professional actions. Following increasing journalistic visits to the Strangeways laboratory in the 1930s, and stung by allegations she was 'about to grow babies in bottles', Honor Fell took a dim view of the popularizing activities she once endorsed and cautioned against sensational portrayals of tissue culture.[5]

The development of organ culture

In 1912, the American anatomists John E. McWhorter and Allen Whipple cultured three-day-old chick embryos *in vitro*. They found these embryos survived for up to 31 hours in a plasma medium and recorded that the brain, nervous and vascular systems developed as normal.[6] In the *Anatomical Record*, they argued that tissue culture allowed scientists to observe developmental phenomenon that had been 'hitherto impossible' with existing histological methods. 'While observing this specimen', they wrote, 'we were impressed with the possibilities that such a method offers for the study of various problems in embryology'.[7] The pair expressed surprise that no other scientist had used tissue culture to study embryonic development since Ross Harrison's 1907 work on the growth of frog nerves *in vitro*. Experimental work, they stated, 'has been confined to the culture of bits of tissue from the embryo or adult animal or from various tumours'.[8] This imbalance was not lost on Harrison himself, who conceded in 1911 that tissue culture's use in developmental studies was minimal when compared to 'phenomena more closely of interest to medicine'.[9]

Thanks largely to the influence of Alexis Carrel, who perceived tissue culture as a potent medical tool, researchers in the 1920s tended to culture small fragments of tissue or tumours to determine what effects various nutrients and chemicals had on growth *in vitro*.[10] This focus led the Russian histologist Alexander Maximow, then working at the University of Chicago, to complain that scientists who employed tissue culture remained 'interested chiefly in general biological problems concerning growth and differentiation of various *cell types* outside the body and the influence of external agents on the structure and function of the living substance'.[11] Deploying an analogy that resonated with the political unrest in his homeland, Maximow continued that practically all research investigated 'anarchistic' behaviour, with scientists studying 'a crowd of various cells without any regular arrangement' (the period between 1918 and 1922, characterized as the Third Russian Revolution, witnessed a series of failed socialist and anarchist rebellions against the Bolshevik Party).[12] Maximow argued this bias was especially surprising, since when it came to the study of development 'the method of tissue culture obviates many experimental difficulties, offering great possibilities and opening a whole new field for investigation'.[13] Culturing embryos and organs, he concluded, 'simplifies the problem to the highest possible level and allows one to observe ontogenetic phenomena under the most favourable conditions'.[14]

Maximow justified these claims by citing McWhorter and Whipple, as well as the Belgian embryologist Albert Brachet who, in 1913, had cultured week-old rabbit embryos in tubes containing plasma clots. Brachet recorded that rabbit embryos survived for up to 40 hours *in vitro* and developed as normal.[15] Maximow also cited the Scottish brothers David and John Gordon Thomson, who used tissue culture during a spell at the Royal Society's Marcus Beck Laboratory in London, and coined the term 'controlled growth' to describe the development of whole organs and embryos *in vitro*. Shortly after arriving at the Marcus Beck Laboratory in 1912, David Thomson had travelled to New York to learn tissue culture at Carrel's Rockefeller laboratory. After returning to London, he and his brother undertook the first documented tissue culture experiments in Britain, cultivating material from chick embryos and surgical patients from the Middlesex Hospital (which is discussed more in the next chapter). Like Carrel and most other scientists who used tissue culture, the Thomsons sought to determine which nutrients facilitated optimal growth of tissue fragments and charted the migration, or 'branching', of cells *in vitro*. But as a control to this work, to see if the presence of a basement membrane impeded cell migration, David Thomson also cultivated a whole toe and a tail end from a nine-day-old chick embryo. After four days, he observed that all these samples had grown significantly and retained their anatomical structure.[16]

After these initial observations, David Thomson detailed how another toe from a nine-day-old embryo grew significantly after five days *in vitro*, and again retained its anatomical form. He also recorded identical observations for an embryonic feather, which grew in culture for a week, and an embryonic eye, which grew for four days. 'In these cases', Thomson claimed, 'the definite shape was retained and no uncontrolled growth occurred from the definite margins of these structures'.[17] He concluded that controlled growth occurred *in vitro* because the uninjured membranes of explanted organs 'form a definite membrane, from which a new uncontrolled growth cannot radiate'. Development only stopped, he continued, when organs became too large and nutrient media 'is unable to penetrate by osmosis into the central portion'.[18] Thomson predicted that continued growth and vitality of explanted parts would be achieved when 'by some means an artificial circulation could be set up so as to enable the nutrient medium to get at the central portions of the tissue'.[19]

Despite their suggestions for technical improvements, the Thomsons did not persist with tissue culture. In 1914, John Gordon Thomson took the position of protozoologist at the London School of Tropical Medicine, while David Thomson left to work on amoebic dysentery at a military station in Gallipoli. Nevertheless, Britain emerged as the leading site for work on 'controlled growth', following Honor Fell's 1923 appointment to the Cambridge Research Hospital. Fell believed that the chick embryo tissue Thomas Strangeways was cultivating for his arthritis research could be used to study embryological development, and suggested they observe the growth of undifferentiated limb buds from early chick embryos. Strangeways and Fell detailed how these samples grew well for up to a week, with histological analysis showing that cartilage nodules formed to a normal and 'comparatively advanced stage of differentiation'.[20] They then cultured eyes from three-day-old embryos, which were much more complex organs than limb buds. Years later, Fell recalled her 'astonishment' at the level of development these eyes underwent *in vitro*: lens fibres formed, pigment rods developed and the retinal epithelium acquired its characteristic structure (although the gross anatomy of each eye was distorted compared to normal development).[21] This seemingly innate capacity for controlled growth strengthened Strangeways's belief that tissues could thrive apart from the body. In his 1926 address to the BMA, he argued the 'surprisingly normal' growth of embryonic parts *in vitro* indicated that 'somatic cells do not require the control of the organism as a whole in order to build up the specific tissues for which they were set apart in life'.[22]

These developmental studies continued after Strangeways's death. Indeed, with Fell installed as director, they became a priority at the renamed Strangeways Research Laboratory. In 1928, Fell demonstrated that the embryonic ear also proceeded to an advanced stage of differentiation *in vitro*.[23] The following year, she collaborated with the biochemist Robert Robison from the Lister Institute, London, to investigate the synthesis of the enzyme phosphatase during bone formation. As part of this latter study, Fell modified her experimental apparatus to permit the cultivation of larger explants for longer periods. In their prior work, she and Strangeways had explanted embryonic parts on the surface of a plasma clot at the bottom of a small centrifuge tube. By being cultivated on top of a large volume of medium, with a large air space above them, the samples had enough nutrient and oxygen to allow growth and differentiation.[24] But these tubes were relatively small and prevented significant development beyond a week. In her work with Robison, which sought to chart the longer term development of bone, Fell explanted samples into a greater volume of medium held in a large watch glass, and placed this at the bottom of a Petri dish with soaked cotton wool to provide a moist chamber.[25] Explanted in this new culture, embryonic bones grew for up to three weeks, more than trebled in size, underwent considerable differentiation and retained their normal anatomical shape.

This 'watch-glass' method promptly became the standard technique for cultivating organs at the Strangeways laboratory, and was adopted by researchers who were interested in studying development *in vitro*. These included the Dutch physiologist Pieter Gaillard, the German embryologist Alfred Glücksmann and a young Cambridge graduate, Conrad Hal Waddington, who was interested in extending Hans Spemann's recent work on embryonic induction.[26] Working with Hilde Rostock at the University of Freiburg, Spemann had discovered that a particular region of the amphibian embryo, which he termed 'the organizer', induced development of complicated structures when transplanted to another embryo.[27] In 1930, Waddington used the watch-glass technique to replicate these results with chick embryos *in vitro* – the first time induction had been observed in warm-blooded embryos.[28] In 1931, he followed this up by culturing whole chick and duck embryos, observing normal development in some and manipulating others to determine which portions gave rise to certain organs.[29] Waddington repeated Brachet's work on rabbit embryos the following year, finding they developed normally for nine days in watch-glass cultures.[30]

In 1933, Waddington began work on the biochemistry of the organizer with the Cambridge embryologists Joseph and Dorothy Needham. By this point, scientists appeared to consider the Strangeways laboratory a site of important embryological research, with tissue culture a useful tool for studying development. At a 1932 meeting of the Society for Experimental Biology, held in Oxford, Fell and Waddington held a session on tissue culture's relevance to experimental embryology. And in February 1934, the Strangeways laboratory was chosen as the site for the inaugural meeting of the 'Embryologist's Club'. As part of this meeting, eight different rooms of the laboratory were used for demonstrations and lectures on the application of organ culture in embryology.[31] In 1937, Waddington and Joseph Needham requested a large grant from the Rockefeller Foundation that would make the Strangeways laboratory the permanent base for their now renowned work on organizer biochemistry (the previous year, they had won the inaugural Albert Brachet Prize, which the Royal Academy of Belgium awarded for the best published work in embryology each year).[32] This application was ultimately unsuccessful. Although Rockefeller trustees expressed admiration for Fell and Waddington, they disagreed with Needham's belief that embryologists should focus solely on biochemical problems, set out in his 1931 book *Chemical Embryology*.[33] The collaboration on organizer biochemistry petered out after the outbreak of World War II, when Waddington was posted to the Royal Navy. Following the end of the War, he left Cambridge to take the chair of animal genetics at the University of Edinburgh. But organ culture remained an important technique at the Strangeways laboratory. During the 1950s and 1960s, Strangeways researchers applied the method to physiological and medical problems – taking advantage of technical improvements, like the use of antibiotics *in vitro*, to culture human as well as animal organs.

Organ culture and 'test-tube babies'

This work on organ culture also featured in popular coverage of tissue culture. Once again, Strangeways researchers were behind much of this popularization. Ronald Canti recorded the organized growth of chick bone on his homemade cinematograph, and included the footage in

'The Cultivation of Living Tissue'. And in her BBC talk on 'The Life of a Cell', Honor Fell claimed that:

> It is an interesting fact that many of the simple, undeveloped tissues of very young embryos will not only grow but will also develop in a culture vessel. For example, if we take a small piece of the soft tissue from the rudimentary jaw of an embryo chick and plant this fragment in a hanging-drop preparation, we shall find that in a few days' time a little piece of bone will begin to form in the culture. By examining the culture under a microscope every few hours the whole course of bone formation can be studied in living tissue.[34]

Julian Huxley also promoted organ culture, writing in the *Manchester Guardian* that it would become 'increasingly valuable in investigating the machinery of normal development'.[35] A 1932 report in *Tit-Bits* was more effusive, describing how Fell and colleagues 'performed the miracle of growing perfect eyes and bones in the laboratory apart from the creatures to which they belonged'. By maintaining embryonic components 'until they were perfectly developed', it continued, 'workers at the Strangeways laboratory have achieved a triumph that may have far-reaching results'.[36]

Rather unusually, *Tit-Bits* did not predict what these 'far-reaching results' might be. But others did, writing speculative reports and fictional stories that envisaged the growth of human babies in tissue culture. These 'test-tube babies' were a regular public presence during the interwar period: appearing in broadsheet and tabloid newspapers, in pulp magazines like *Amazing Stories*, highbrow periodicals like F.R. Leavis's *Scrutiny*, in speculative essays, popular novels and even in scientific journals like *Nature*. This profile was testament to the way that test-tube babies resonated with several interwar concerns. The prospect of growing babies in tissue culture symbolized the belief that biologists were attaining unprecedented control over natural phenomena. At the same time, several writers used test-tube babies to link debates on biological control to concerns over industrial culture, framing test-tube babies as little more than another assembly-line commodity. Others, meanwhile, linked test-tube babies to the interwar preoccupation with reproductive control, population health and the changing social status of women. Advocates of eugenics and birth control claimed test-tube babies would overcome degeneration and free women from the rigours of pregnancy and childbirth. On the other hand, writers opposed to birth control, political suffrage and eugenics argued they would subordinate men and further the scientific interference in human affairs.

These varying standpoints were lent credibility by the belief that test-tube babies were imminent. As *Tit-Bits* declared in 1938, they were 'what science looks like producing next'.[37] Such claims invariably drew upon the development of organ culture methods. This was evident in the first account of *in vitro* reproduction, *Daedalus, or Science and the Future*, written by the Cambridge geneticist J.B.S. Haldane. Haldane first drafted this essay in 1912, while he was an undergraduate at Oxford. He returned to it after serving on the Front Line in 1915, recalling chemical attacks and men 'running, with mad terror in their eyes, from gigantic steel slugs, which were deliberately, relentlessly, and successfully pursuing them'.[38] After Haldane read it to a meeting of the Cambridge Heretics in 1923, the publisher Kegan Paul released *Daedalus* as the inaugural pamphlet of its *To-day and To-morrow* series. Indicating the demand for futurological speculation in the interwar years, the series published over 50 essays on the future of issues like science, technology, sport, morality, the body, women and art between 1924 and 1931.[39] Some topics were the subject of more than one book, with different authors positing alternate futures. Science, and biology in particular, were especially popular themes. Following Haldane, the crystallographer J.D. Bernal and the zoologist H.S. Jennings, amongst others, explored the future of the body, eugenics, Darwinism, evolutionary psychology and genetic disease. This focus again reflected the view that biology was becoming the most powerful of the sciences. But it was also influenced by the success of *Daedalus*, which sold over 15,000 copies, went through seven impressions in three years and was praised in the *Observer* as 'brilliant and audacious'.[40]

Taking the premise that 'the centre of scientific interest lies in biology', Haldane used *Daedalus* to outline 'a few obvious developments which seem probable in the future state of biological science'.[41] Foremost amongst these was the development of what he called '*ectogenesis*', a term that originated in the 1880s to denote the growth of bacteria outside the body. Haldane extended this meaning to describe a method whereby:

> we can take an ovary from a woman, and keep it growing in a suitable fluid for as long as twenty years, producing a fresh ovum each month, of which 90 per cent can be fertilized, and the embryos

grown successfully for nine months, and then brought out into the air.[42]

Haldane predicted the first 'ectogenetic child' in 1951. By the 1960s, he claimed, over 60,000 children would be 'produced' this way, and by the 21st century less than 30 per cent of British children would be 'born of woman'.[43]

Although *Daedalus* was set thousands of years in the future, Haldane firmly linked ectogenesis to interwar concerns surrounding eugenics and birth control. During the early 1920s, British eugenicists lamented the loss of many 'superior' young men in World War I, a declining birth rate among the upper and middle classes, and the proliferation of 'degenerate' working class stock.[44] As he turned toward leftist politics at the end of the 1920s, Haldane criticized mainline eugenics for its emphasis on preventing the seemingly degenerate from reproducing, and argued that population health could only be improved by raising living and educational standards for the working classes.[45] But perhaps because he began it when he was an undergraduate, and a member of the Oxford Eugenics Society, *Daedalus* expressed no such misgivings.[46] Indeed, Haldane's futuristic narrator presented ectogenesis as a eugenic panacea, recounting how:

> The small proportion of men and women who are selected as ancestors for the next generation are so undoubtedly superior to the average that the advance in each generation in any single respect, from the increased output of first-class music to the decreased conviction for theft, is very startling. Had it not been for ectogenesis there would be little doubt that civilization would have collapsed within a measurable time owing to the greater fertility of the less desirable members of the population in almost all countries.[47]

The 1920s also saw increasing synergy between the birth control, feminist and eugenics movements.[48] Haldane furthered this association in *Daedalus*, presenting ectogenesis as a boon to campaigners like Marie Stopes, who argued that contraception would increase women's sexual freedom by removing the spectre of pregnancy, and would correct population health by preventing 'wastrels' from breeding.[49] Haldane predicted that by completely separating 'sexual love and reproduction', ectogenesis would create a utopia where 'mankind will be free in an altogether new sense'.[50]

Haldane also linked *Daedalus* to contemporary science, reminding his readers that 'not one of the practical advances which I have predicted is not already fore-shadowed by recent scientific work'.[51] To justify his account of ectogenesis he detailed how Albert Brachet had 'grown rabbit embryos in serum for some days' in 1913.[52] From this, he followed, it was a short step to cultivating rats, pigs and 'the first ectogenetic child'.[53] Reviews of *Daedalus* did likewise. As *Nature* claimed, Haldane's account of ectogenesis did not appear far-fetched 'if what has already been done with tissue culture is remembered'.[54] This association was further strengthened during Thomas Strangeways's final lecture course in December 1926. Although he never published the results, Strangeways had extended his developmental studies by culturing whole chick embryos.[55] Detailing this experiment at the beginning of the first lecture, he told his students how 'by a special technique the unincubated blastoderm may be cultivated entire and will continue to grow and develop for several days'.[56] Significantly, however, Strangeways did not outline any scientific implications this work may have held. Instead, he declared that: *'It will thus be seen that the idea of the "test-tube baby" is not inherently impossible'.*[57]

This claim again demonstrates how Strangeways constructed tissue culture in line with broader interwar concerns. It is also significant for another two reasons. Firstly, it represents the first endorsement of ectogenesis from someone who had actually maintained embryos *in vitro*. Secondly, it predates the first documented use of the term 'test-tube baby' by nearly ten years: as it stands, the *Oxford English Dictionary* assigns its origins to a 1935 United States debate on artificial insemination. But unlike his 'revival' of sausage meat, no newspaper or scientific journal covered Strangeways's reference to test-tube babies. The term did not reappear in Britain until 1937, when a *Daily Mirror* report on biological research announced: 'This is not a prophecy – It's news about Test-Tube Babies!'[58] But cultured babies remained a regular public presence in the interim, with newspapers generally referring to them as 'ectogenetic children' or 'chemical babies'.

Like Haldane, other authors presented ectogenesis as a solution to issues surrounding birth control, feminism and population health. In his 1930 essay *Chronos, or the Future of the Family*, the physician Eden Paul aligned it with campaigns for 'the emancipation of women' and stated 'there is no sex reformer but must wish that woman could be freed from the slavery of child-bearing'.[59] The same year, in *The World in 2030 A.D.*, the Conservative politician Lord Birkenhead predicted that ectogenesis would 'vitally transform the status of women in society' by liberating them 'from the dangers of childbirth'.[60] Birkenhead also believed ectogenesis had positive eugenic implications, allowing society to 'produce the human types it most needs, instead of being forced to absorb all the unsuitable types which happen to be born'.[61] Like Haldane, who protested at overt similarities to *Daedalus*, Birkenhead justified his predictions by outlining how 'the foetus of various species has been removed from the maternal organism and further developed by skillful manipulation in biological laboratories'.[62]

Other writers incorporated organ culture into more radical accounts of reformed bodies, imagining the *in vitro* production of monstrous animals and human-machine hybrids. This prospect underpinned Julian Huxley's only foray into fictional writing, 'The Tissue Culture King'. First published in the *Cornhill Magazine*, which was edited by his father, this short story later reappeared in a 1927 edition of *Amazing Stories*. Borrowing heavily from Joseph Conrad's *Heart of Darkness* and H.G. Wells's *Island of Doctor Moreau*, 'The Tissue Culture King' centred on a group of African explorers who stumble into an unnamed country where a British scientist, Dr. Hascombe, has attained a powerful position thanks to his prowess with *in vitro* methods. The explorers learn that Hascombe has profited from local veneration of the tribal King by culturing portions of his tissue and propagating them 'indefinitely, to ensure their protecting power should reside everywhere in the country'.[63] He has also seized on the tribal 'fancy for the grotesque in animals' and grafted between cultured embryos to produce three-headed snakes and two-headed toads. Here, Huxley made explicit the link between tissue culture and industrial mass-production techniques that other writers, including his brother Aldous, later explored in detail. Hascombe tells the exploring party that tissue culture allowed him to apply 'the mass production methods of Mr. Ford' to breeding, resulting in the large-scale creation of 'double-headed and cyclopean monsters'.[64]

The possibility for a radical reconfiguration of the body via tissue culture was explored, and celebrated, in greater detail by J.D. Bernal's 1929 essay *The World, The Flesh and the Devil*, which also appeared in the *To-day and To-morrow* series. Here, Bernal outlined various ways that humanity could scientifically direct its own evolution in the future: suppressing the superego through Freudian psychology, using sophisticated engineering to conquer space and utilizing new biological techniques to augment the body. Drawing from Haldane, Bernal predicted that human life would start 'in an ectogenetic factory'.[65] He also argued that the colonization of space would be far easier if the limitations of the body were overcome, since scientists would not have to worry about providing oxygen or water. Bernal predicted that scientists would therefore have to undertake a profound 'mechanization of the body' in order to leave the Earth, claiming that 'man must actively interfere in his own making and interfere with it in a highly unnatural manner'.[66] This mechanization entailed dispensing with 'the useless parts of the body' and replacing them with more efficient devices, including artificial limbs and heightened sensory organs.[67]

While Bernal admitted it was 'difficult to form a picture of the final state' of this mechanized body, as a 'number of typical forms would be developed, each specialized in certain directions', he nevertheless detailed 'what might be called its first stage'.[68] No trace of the human form remained. In its place, Bernal presented a system of interconnected chambers containing vital organs, with these *in vitro* parts maintained thanks to an artificial circulatory system:

> Instead of the present body structure we should have the whole framework of some very rigid material, probably not metal but one of the new fibrous substances. In shape it will be rather a short cylinder. Inside the cylinder, and supported very carefully to prevent shock, is the brain with its nerve connections, immersed in a liquid of the nature of cerebro-spinal fluid, kept circulating over it at uniform temperature. The brain and nerve cells are kept supplied with fresh oxygenated blood and drained of de-oxygenated blood through their arteries and veins which connect outside the cylinder to the artificial heart-lung digestive system – an elaborate, automatic contrivance.[69]

At the back of this cylinder, Bernal continued, were the eyes and ears, enhanced by connection to projecting telescopes, microscopes, microphones and wireless attachments. Completing the system was a 'locomotor apparatus of different kinds, which could be used for slow movement, equivalent to walking, for rapid transit and for flight'.[70]

Like Haldane and Birkenhead, Bernal presented these changes as a significant advance, creating the 'perfect man such as that the doctors, the eugenicists and the public health officers hope to make of humanity'.[71] Enabling humans to colonize space, and crafted by new scientific techniques, this *in vitro* body offered a celebration of both the aspirations and technical aptitude of modern science. In this respect, *The World, The Flesh and the Devil* echoed Futurist writing and art from the 1910s and 1920s, which glorified 'man multiplied by the machine' (though Bernal eschewed the overt militarism and fascist overtones of much Futurist work). According to the Italian writer Filippo Thomasso Marinetti, whose 'Founding and Manifesto of Futurism' was published by *Le Figaro* in 1909, the fusion of flesh and technology would create a superior 'non-human type', endowed with new organs and sensory capabilities that enabled it to withstand 'a world of ceaseless shocks'.[72] But while Marinetti and other Futurists saw the 'non-human type' as the product of interaction with new automobiles and planes, Bernal claimed the 'radical alteration of the body' lay in the application of new biological and surgical methods.[73] Although he did not cite organ culture, it certainly appeared to inform *The World, The Flesh and the Devil*. His reference to ectogenesis was a clear nod to organ culture, while the detailed account of the apparatus needed to sustain whole organs, comprising glass cylinders and nutrient medium, was reminiscent of the methods the Thomsons, Strangeways and Fell employed to study 'controlled growth' *in vitro*.

But not all views of the future were so positive. Several of the *To-day and To-morrow* essays framed scientific progress in distinctly ambivalent terms. In *Icarus, or the Future of Science*, written as a retort to the utopian *Daedalus*, the philosopher Bertrand Russell claimed science would more likely be used as a tool to consolidate the power of dominant groups. 'Men sometimes speak as though the progress of science must be a boon to mankind', he wrote, 'but that, I fear, is one of the more comfortable delusions which our more disillusioned age must discard'.[74] Others made ectogenesis the focus of their dystopian predictions. In his 1924 essay *Lysistrata, or Woman's Future and Future Woman*, the right-wing writer Anthony Ludovici predicted that a feminist elite would use ectogenesis to control their own reproduction and, in the process, subordinate men.

Ludovici was a staunch anti-feminist, who opposed the 1918 decision to give the vote to women over 30, and declared in 1921 that 'nothing good can come of teaching women anything'.[75] He used *Lysistrata* to criticize political suffrage and campaigns for greater female access to birth control, imagining a future where women displaced men from many vital roles, including reproduction. Throughout, he railed against the 'body despising values' that underpinned the development of technological replacements for masculine roles and actions. Ludovici argued that these 'body despising values' stemmed from scientific and feminist collusion, and predicted they would foster a 'marked decline in the ability, versatility and masculinity of men'.[76] He outlined how modern technologies would reduce men to 'little more than machine-minders or adding-machines, exercising few if any of their highest faculties'.[77] Ectogenesis offered the culmination of this process, displacing men from the very act that Ludovici saw as constitutive of masculinity. He forecast that in the near future, where machines replaced men in factories and on the battlefield, and women replaced men in Parliament, 'the whole act of fertilization will be consummated in the surgery, just as vaccination is now'.[78] Ludovici concluded by warning that once this 'Feminist ideal of a complete emancipation from the thraldom of sex' was attained, 'men will be frankly regarded as quite superfluous'.[79]

With ectogenesis as the norm, and controlled by a 'Parliament of women', Ludovici envisaged that traditional procreation would be punishable by castration and 'men of vigorous sexuality will be elim-inated in order to make way for a generation of low-sexed, meek, and sequatious lackeys'.[80] In this emasculated future men appeared no more than 'a source of irritation and indignation'. Women limited male births to the half a percent of the population necessary for sperm supply, and slaughtered adult men they deemed surplus to require-ments.[81] Were this not dystopian enough, Ludovici believed that fem-inist civilization would petrify and ultimately collapse, 'as it is not in woman's nature to be inventive or make great discoveries'.[82] Through-out, and in a familiar move, he legitimated his predictions by arguing that ectogenesis was 'all potential in the scientific achievements of our day' and a likely outcome of 'success on tissue culture'.[83] Like Haldane, Birkenhead and Strangeways, Ludovici claimed that the cultivation of animal embryos foreshadowed the growth of human babies in tissue cul-ture. Indeed, he claimed that the contemporary obsession with eugenics and birth control made this progression inevitable, since no scientist 'will be allowed no rest until a technique is developed that meets the public demand'.[84]

Tissue culture was central to a similarly foreboding, though less misogynist, vision of the future in the philosopher Olaf Stapledon's 1929 novel *Last and First Men*. This was perhaps the most sweep-ing futurological work of the interwar period, charting the rise and fall of 18 distinct human races and civilizations over two billion years.[85] As Stapledon set out in a preface, his predictions were under-pinned by the belief that 'any attempt to conceive such a drama must take into account whatever contemporary science has to say'.[86] His extrapolations from actual research were widely seen as one of the book's strengths. In a congratulatory letter, J.B.S. Haldane described Stapledon's treatment of science as 'unimpeachable', and wrongly assumed that the author must have been a research scientist.[87]

Stapledon made tissue culture fundamental to the creation of the so-called 'Fourth Men', which scientists created by fertilizing human ovaries *in vitro*.[88] In a chapter entitled 'Man Remakes Himself', he detailed how scientists manipulated the resultant embryos *in vitro* to fashion 'an organism which consisted of a brain twelve feet across, and a body most of which was reduced to a mere vestige upon the under-surface of the brain'.[89] This 'fourth man' was sustained by complicated machinery and bore no resemblance to its human creators:

> This fantastic organism was generated and matured in a building designed to house both it and the complicated machinery which was necessary to keep it alive. A self-regulating pump, electrically-driven, served it as heart. A chemical factory poured the necessary materials into its blood and removed waste products, thus taking the place of digestive organs. Its lungs consisted of a great room full of oxidizing tubes, through which a constant wind was driven by an electric fan.[90]

Following the premature death of this 'fourth man' due to the weight of its own brain, Stapledon documented how scientists worked on improvements: taking 400 years to successfully develop a 'great brain' that was housed in a 'turret of ferro-concrete some forty feet in diameter'.[91]

On the surface, this cultured brain appeared homologous to the 'mechanized men' described in *The World, The Flesh and the Devil*. But unlike Bernal, Stapledon used the 'fourth men' to pass comment on the negative aspects of mechanization and industrial culture. A lifelong socialist, he shared the Marxist view that machines disenfranchised and subordinated human workers. As he claimed at its outset, *Last and First Men* offered a critique of the 'cruder aspects of Americanism', including the mass-production of commodities and the regimented management of workers' bodies.[92] Stapledon's 'great brain' clearly reflected these views. Due to its enormous intelligence, this product of human labour quickly became an 'autocrat of the state' and forced scientists to mass-produce more its kind. Soon, Stapledon detailed, these fourth men colonized the world and lived 'in the tropics, in the Arctic, in the forests, the deserts, and on the ocean floor'.[93] Following a brief uprising, they eventually destroyed the human race that had developed and propagated them.

By presenting a scenario where mass-produced creatures sub-ordinated their human manufacturers, *Last and First Men* chimed with the belief that humans were increasingly enslaved to the regimented demands of modern industrial culture. Stapledon's narrative was similar to Karel Capek's 1921 play *Rossum's Universal Robots (R.U.R.)*, which documented how mass-produced 'robots' replaced skilled trades-people and ultimately turned on their human creators.[94] It also echoed Fritz Lang's 1927 film *Metropolis*, where workers ceaselessly toiled on machines to power the industrial city and were duped into a suicidal uprising by the artificial robot 'Maria'.[95] While these gloomy visions of the future all conveyed the same distrust of machines and mass-production, each presented different techniques as the means by which artificial beings were produced and humans subordinated. Capek's robots were biological composites of different organs, but *R.U.R.* did not outline how they were assembled; the robot Maria in *Metropolis*, meanwhile, was a metallic entity powered by electricity. *Last and First Men*, on the other hand, was one of several accounts that presented organ culture as the route to an enslaved and dark future.

In 1927, *Amazing Stories* published 'The Machine Man of Ardathia' by Francis Flagg, a pseudonym for the American novelist George Henry Weiss.[96] This brief story involved an encounter between a present-day American and a time-travelling humanoid from 28,000 years in the future. The time-traveller is suspended in a cylinder by 'an intricate arrangement of glass and metal tubes', and tells his companion that humanity evolved to this state when it realized 'bodily advancement lay in, and through, the machine'.[97] This future race, the 'Ardathians', develop from a fertilized cell *in vitro* and spend their whole lives encased in culture vessels, incorporating machinery into their bodies as they grow (see Figure 3.1). Weiss's Ardathian tells his twentieth century companion that the development of tissue culture marked the 'starting point from which you may be able to follow my explanation of man's evolution from your time to mine'.[98] But while the present-day human has 'heard tell of chicken hearts being kept alive in special containers', he is horrified by the method's implications. Like Stapledon's fourth men, Weiss framed the Ardathian as emblematic of the dehumanizing effects of machine culture: unable to leave its cylindrical home, it is incapable of empathy and disgusted by the prospect of human con-tact. What is more, its reliance on technological prosthetics appears to have fostered physical degeneration. The Ardathian is sexless, resem-bles a malformed embryo and is a mere three-foot tall, with 'skinny, flabby' legs and arms 'more like short tentacles'.[99] The fusion of flesh

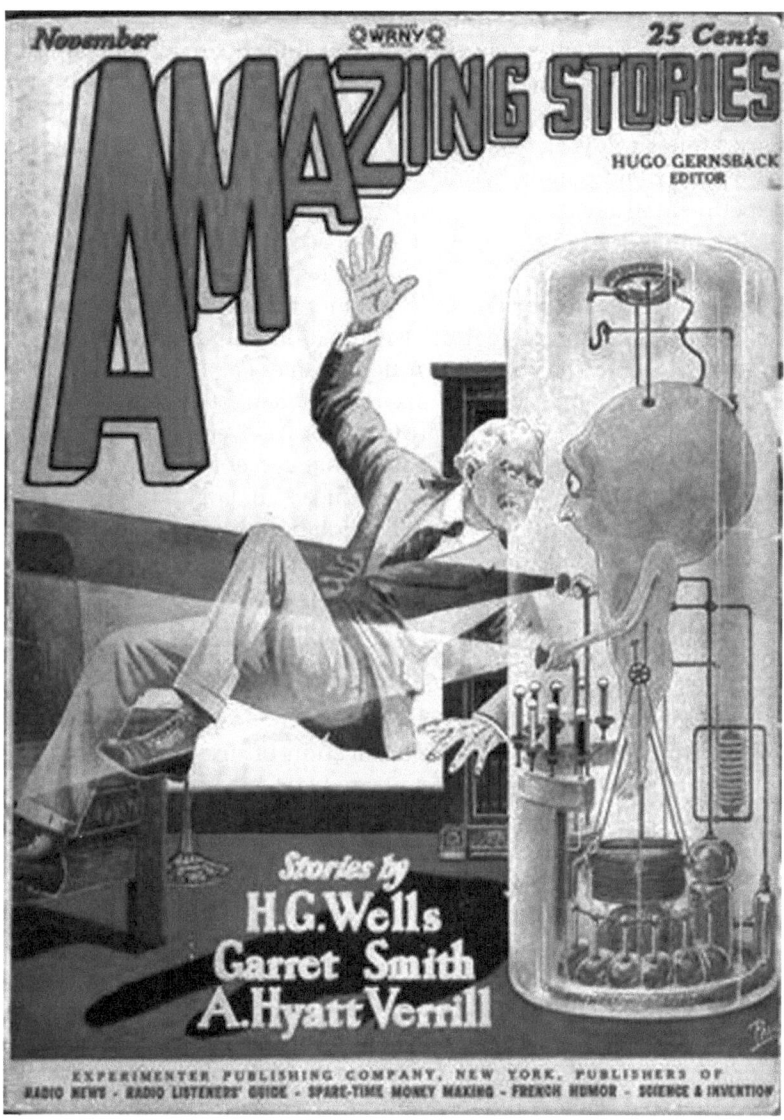

Figure 3.1   Cover to a 1927 edition of *Amazing Stories*, showing the 'Machine Man of Ardathia'. Illustration by Frank R. Paul, used with permission of the Frank R. Paul estate.

and technology here is a world away from J.D. Bernal's 'perfect man' or the Futurist celebration of 'man multiplied by the machine'. Indeed, Weiss firmly presents the Ardathian's account of ectogenesis and mechanical life as 'an appalling story'.[100]

The following year, *Amazing Stories* published a short story by the psychiatrist and popular writer David H. Keller, entitled 'A Biological Experiment'. Here, ectogenesis was again presented as the cornerstone of an oppressive future society. Keller's account begins in the 1920s, when the development of tissue culture led to 'the discovery that the human ovary could be kept alive' *in vitro*. He then outlined how the deep-seated fear of biological degeneration impelled scientists to produce eugenically perfect 'synthetic babies'.[101] By the year 3928, when most of the story is set, sex is forbidden, sterilization is compulsory, and couples have to apply to a central Government Ministry for 'parent permits'. The synthetic babies that successful couples acquire are standard commodities. Mass-produced in large culture vessels (see Figure 3.2) and homogenized by radiation treatment, they are no different to the radios the Government distributes to ensure uniformity of thought. This circulation of standardized children and displacement from the reproductive process offered a potent criticism of eugenics, mass-production and the notion of a scientific utopia advanced in books like *Daedalus*. In Keller's future: 'No-one had a right to a private opinion; everyone had to think like everyone else; there was a gradual death of individuality'.[102] Despite the eradication of disease, degeneration and social problems, his characters are beset by a pervasive sense of alienation. As one declares: 'In spite of the perfection no one is really happy!'[103] Pursuing this theme, Keller concluded that unfettered scientific and industrial progress deprived people of their creativity, individuality and, indeed, their basic humanity. It is struggle, death and love, he argued, 'that made existence human'.[104]

An identical moral underpinned Aldous Huxley's 1932 novel *Brave New World*, which broadened the futurological preoccupation with eugenics, mass-production and sex reform to include behavioural conditioning, drugs, the cinema and consumer culture. A biting satire of the belief that science and mechanization guaranteed social progress, *Brave New World* was undoubtedly the most influential dystopia of the interwar years, and it still informs present-day discussion of cloning, genetic engineering and embryo research.[105] But the tendency to align *Brave New World* with our current concerns obscures just how much it drew on interwar books like *Daedalus* and *Icarus*, and brought together Huxley's existing views on utopia, science and industry. In his 1923

Figure 3.2 *Amazing Stories* illustration to 'A Biological Experiment', showing the mass-production of 'synthetic babies'. Illustration by Frank R. Paul, used with permission of the Frank R. Paul estate.

novel *Antic Hay*, Huxley first critiqued visions of a utopian future as 'too late in the day', and out of synch with a world coming to terms with the carnage of war and political unrest across Europe.[106] *Antic Hay* also satirized the biological obsession with controlling life through the bumbling physiologist Shearwater, who is too preoccupied with his experiments to notice the parade of men bedding his wife.

Shearwater was modelled on J.B.S. Haldane, who was a childhood friend of Julian and Aldous Huxley.[107] Like Haldane and his brother, Aldous Huxley believed biology had attained unprecedented power since the turn of the century, but he was more inclined to dwell on its potential for misuse. In his contribution to a 1933 volume on *Science and the Changing World*, he combined this pessimism with his disdain for modern industrial culture and argued, like Russell, that:

> Science is morally neutral; it comes to good or evil according as it is applied. Ideally, science should be applied by humanists. In this case it would be good. In actual fact it is more likely to be applied by economists, and so to turn out, if not wholly bad, at any rate as a mixed blessing.[108]

Science, Huxley followed, was a 'heaven-sent instrument' for economists and industrialists. It offered the perfect route to social conformity by ensuring stability and productivity, creating a 'race not of perfect human beings, but of perfect mass-producers and mass-consumers'.[109]

Huxley expanded on this in *Brave New World*, imagining a highly regimented world 600 years in the future, where biology, psychology, genetics and medicine are intertwined with 'the dreadful religion of the machine' in order to suppress individuality and maintain social order.[110] Unlike Keller and Ludovici's dystopias, the inhabitants of this world are blissfully unaware of their subordination. They are induced to 'like their inescapable social destiny' as children by Pavlovian conditioning and hypnopaedia, and their docility is maintained through adulthood thanks to the numbing drug *soma*, the vicarious pleasures of cinematic 'feelies' and leisure pursuits such as Obstacle Golf.[111] They are also conditioned to literally worship mechanization. Conventional religion has been replaced by veneration of Henry Ford, and year zero of this new era is dated to the production of his first Model-T car.

Thanks to ectogenesis, moreover, the population are industrial products themselves, fashioned on production lines like other consumer goods. As the Director of the London Hatchery tells a group of students, one of the 'major instruments of social stability' involves the scientific production of human embryos – 'the principle of mass-production at last applied to biology'.[112] As part of ectogenesis, or 'Bokanovsky's Process', X-rays force *in vitro* ovaries to asexually produce thousands of embryos. Cultured in a bed of peritoneum and stimulated to grow by regular doses of thyroxin, these identical embryos then pass down a conveyor belt for 267 days, where low-grade workers prepare them for their predestined roles by exposing them to heat, starving them of oxygen and rendering them infertile.[113] The only character horrified by this 'nightmare of swarming indistinguishable sameness' is the outsider John, born naturally in a Savage Reservation and brought to London by an ambitious scientist.[114] Like the characters in 'A Biological Experiment', John argues that human life is meaningless without danger, suffering and freedom. Claiming the 'right to grow old and ugly and impotent; the right to have too little to eat; the right to be lousy; the right to live in constant apprehension of what might happen tomorrow', he leaves London and seeks refuge in an abandoned lighthouse, where, pursued by sightseers, he eventually hangs himself.[115]

Huxley did not cite contemporary organ culture research in *Brave New World*, but a favourable review by Joseph Needham made the connection explicit. Writing in F.R. Leavis's *Scrutiny* magazine, Needham claimed that many readers of *Brave New World* would doubtless 'say we can't believe all this, the biology is all wrong, it couldn't possibly happen'.[116] If this were the case, he argued, then only biologists would appreciate the 'full force' of Huxley's predictions:

> Unfortunately, what gives the biologist a sardonic smile as he reads it, is the fact that *the biology is perfectly right*, and Mr. Huxley has included nothing in his book but what might be regarded as legitimate extrapolations from knowledge and power that we already have. Successful experiments are even now being made in the cultivation of small mammals in vitro, and some of the most horrible of Mr. Huxley's predictions, the production of numerous low-grade workers of precisely identical genetic constitution from one egg, is perfectly possible.[117]

Needham's reference to 'successful experiments' in the cultivation of embryos was almost certainly a nod to Conrad Waddington's research. At a London meeting of the Society for Experimental Biology in April 1932, Needham had heard Waddington lecture on 'The Tissue Culture

Technique in Early Embryology', and took detailed notes on the cultivation of chick, duck and rabbit embryos.[118]

Although Needham did not name the scientist or institution that
was working toward 'the most horrible of Mr. Huxley's predictions',
journalists were nevertheless making the connection with the Strangeways laboratory. As outlined in the introduction, in February 1935
Honor Fell wrote to Henry Dale regarding a visit from a *Sunday Dispatch*
journalist, who was convinced Strangeways scientists 'are about to
culture babies in bottles'.[119] Tellingly, she told Dale the journalist had
refused to believe her denials 'in view of information he has received
from elsewhere'.[120] Replying the next day, Dale urged Fell to 'refuse
to see, or to communicate with, any newspaper reporter under any
circumstances'.[121] He suggested 'the only remedy' may be to sue the
Dispatch, though he admitted that 'a newspaper with vast resources
behind it would probably welcome an expensive libel action, as providing
then with further good copy'.[122]

The *Sunday Dispatch* never ran its story on 'test-tube babies', but the
encounter with its journalist certainly affected Honor Fell. Later in
1935, she wrote an article in the *British Journal of Radiology* that highlighted her frustration at scientists who sensationalized tissue culture.
Fell criticized the 'many extravagant claims' that scientists made regarding tissue culture, which 'has sometimes impelled the more critically
minded to regard the whole subject with suspicion'.[123] She concluded
that tissue culture should be framed as 'merely a valuable technique
with peculiar advantages and limitations ... only valuable when applied
by adequately qualified people to suitable problems'.[124]

But Fell's plea ultimately came to nothing. Newspapers increasingly
linked the Strangeways laboratory to ectogenesis, and often drew upon
scientific claims in doing so. In 1936, for example, the *Daily Express*
published a long article on the development of organ culture in Cambridge, detailing how 'beneath microscopes ... living tissues are growing and developing exactly as they would in the complete parts of
living animal bodies'.[125] The report portrayed Fell as a clearly reticent
figure, who 'smokes a little, smiles a little, and talks less'. Her main
contribution was to reassert that tissue culture was simply 'valuable
in attempts to analyse complicated processes which occur in the
human and animal body'.[126] But this was immediately contradicted by
'an unnamed scientist from another Cambridge lab', who told the
Express that Fell's research did indeed mark 'the first steps to the *Brave
New World* visualized by Aldous Huxley, with babies cultivated in
test tubes'.[127]

Other articles followed the same pattern. In 1937, the *Daily Mirror* covered the work of the Strangeways scientist Petar Matrinovich, who studied the *in vitro* development of rat ovaries. The report noted how these explants 'remained healthy after twenty-two days', but admitted Matrinovich 'does not talk about bottle babies'. Nevertheless, it claimed that his cultivation of mammalian ovaries ensured that 'science moves slowly in their general direction'.[128] The following year, a long *Tit-Bits* report on tissue culture predicted that scientists would cultivate human babies 'within a few years'.[129] Of all the press articles that followed *Brave New World*, this was the most ambivalent. Its headline questioned 'Could You *Love* a Chemical Baby?', and the report went on to frame tissue culture as the route to 'sexless, soulless creatures of chemistry'. Echoing Huxley, Stapledon and Keller, amongst others, it ended by asking whether these 'chemical babies' would 'conquer the true human beings'.[130] The future it envisaged involved an 'end to humanity' – the ultimate dystopia.

Like other newspapers, *Tit-Bits* linked ectogenesis to Honor Fell and the Strangeways laboratory, 'where the new technique of tissue culture is carried out'.[131] There is good evidence that this prevalent association changed Fell's views on popularization. In a 1939 letter to the physiologist A.V. Hill, then Biology Secretary at the Royal Society, she requested an increase in funding for Conrad Waddington, whose annual stipend had recently fallen from £800 to £500. Fell told Hill that Waddington was supplementing his income by writing for popular magazines like *Discovery* and the *Listener*, which she dismissed as a 'great waste of

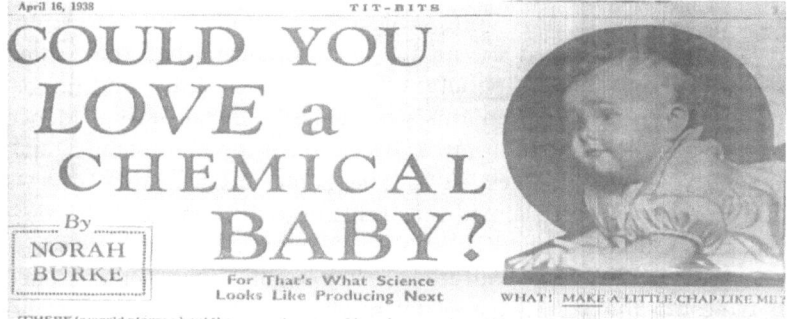

Figure 3.3   Tit-Bits claims 'chemical babies' are imminent, and confronts its readers with a difficult question. Courtesy of the Wellcome Library for the History of Medicine.

time'.[132] Indeed, the express purpose of her letter was to secure money so he could 'abandon these activities'.[133] This represented a notable about-turn from the woman who once endorsed 'propaganda', gave BBC radio lectures, wrote for the *Listener* herself and even planned a popular book on tissue culture. Although Fell continued to lecture to schoolchildren and students, as she would throughout her career, she now spent time dismissing the 'extravagant claims about tissue culture, which actual experience in the laboratory fails to justify'.[134]

The increasingly sinister representations of tissue culture appear to have brought about this shift, as Fell's popular activities from the early 1930s carried no such warnings. In particular, Fell appeared to be most perturbed by the widespread predictions of test-tube babies, which she confessed to Henry Dale caused her 'considerable anxiety'.[135] This is understandable, since several newspapers portrayed test-tube babies as a likely outcome of her own research on organ culture. When Fell had been happy to popularize tissue culture in the early 1930s, her letters to Malcolm Donaldson show that she did so in order to foster popular confidence in the method and her laboratory. Reports that portrayed her as the likely creator of 'soulless' chemical babies were clearly not the sort of propaganda that she had in mind.[136]

Conclusions

Honor Fell's staunch rejection of the 'extravagant claims' surrounding tissue culture may appear to vindicate a number of historical assumptions. Firstly, it lends support to the belief that interwar tabloids and pulp magazines were rife with sensationalism, and that intellectuals looked upon these publications with disdain.[137] In particular, historians have long believed that most interwar scientists were so horrified by media sensationalism that they shied away from any form of popular engagement. Secondly, and importantly, Fell's behaviour in this period has been taken as evidence of a longstanding dichotomy between scientific and popular attitudes to work on tissues. In a 1998 *Lancet* article, Lori Andrews and Dorothy Nelkin claimed that Fell's refusal to engage with journalists in this period highlighted 'the gap between scientific and social perspectives' on tissue culture.[138] They argued that her reticence is typical of the 'dismissive or defensive reactions' of scientists who are believed to work on human tissue – ignorant of, and juxtaposed to, the 'cultural values and social concerns' expressed in popular outlets.[139]

But these assumptions are misleading. Peter Bowler has recently challenged the myth of the isolated interwar scientist, showing that a significant proportion of British scientists regularly engaged with popular audiences in the interwar period.[140] My last two chapters reinforce Bowler's assessment, demonstrating how several scientists popularized tissue culture during the 1920s and 1930s. Many of these figures, such as Julian Huxley, Haldane, Needham and Waddington, are already well-known 'public scientists'. Others, including Honor Fell and Ronald Canti, are less familiar names, but their radio talks, popular articles and cinematic films played a large part in making tissue culture a publicly visible technique during the interwar years. What is more, the media interest that caused Honor Fell's reticence, which Nelkin and Andrews cite as evidence of a dichotomy between scientists and the public, resulted from a dynamic interplay between professional and popular concerns. Far from being the product of a sensationalist or misinformed tabloid mindset, opposed to scientific perspectives, test-tube babies were a product of a scientific engagement with broader debates on eugenics, mass-production and birth control. Figures like J.B.S. Haldane, Thomas Strangeways and Joseph Needham publicly discussed them well before the *Daily Express*, the *Daily Mirror* or *Tit-Bits*. Honor Fell was clearly aware of this, criticizing her fellow scientists for the exaggeration surrounding tissue culture. And this exasperation merely increased when scientific endorsement of ectogenesis rebounded in the form of increasingly negative newspaper stories on 'chemical babies'. Far from revealing a dichotomy between scientists and the public, Honor Fell's attempts to dampen media interest in test-tube babies show instead how public images often arise thanks to scientific practices and claims, and can interact with and shape science itself.

4
Converting Human Material into Tissue Culture, *c*.1910–70

The numerous accounts of 'test-tube babies' that appeared in the inter-war period documented the cultivation of human material. But although these stories drew upon biological practices and were endorsed by scientists, human tissue was in fact hard to maintain and was used far less than chick tissue. Human material rarely survived for more than a few days *in vitro*, and generated little results of scientific or medical interest. But researchers continued to obtain and culture various types of human tissues, as they believed that any findings obtained from them would be more applicable to patients.

Human tissue became far easier to culture after 1945, however, thanks to improvements in culture media and the deployment of antibiotics *in vitro*. Like tissue culture in general, human tissue culture was now perceived less as an esoteric, contested technique and more as a standardized and essential procedure – crucial to work with distinct medical relevance, such as vaccine development. The cells and tissues that were cultured, in turn, went from being represented as the source of possible catastrophe to 'complex chemical workshops' that were the basis of many potential therapies.[1]

Tissue culture continued to feature in newspaper reports, but was now portrayed in a positive light. Amidst celebratory postwar coverage of scientific and medical progress, journalists approvingly detailed the use of human tissues in the fight against disease. Importantly, none sounded unease at the fact that human tissues were increasingly cultured 'outside of the body of which they were once part'.[2] This undermines claims that the scientific acquisition and use of human material invariably meets popular opposition. Excised tissues were readily exchanged between clinicians and scientists for most of the twentieth century, without any thought for patient consent or ownership, because their collection and use in research was widely viewed as unproblematic.

Standing at the shoulder of surgeons: Acquiring human tissue *c.*1910–45

Several historians have detailed how the growing experimental ethos in biology led to increased collection of human and animal materials, as scientists established greater links with animal breeders, slaughterhouses and nearby hospitals.[3] When it came to human tissue, prominent examples include the embryologist Wilhelm His, and his American successors Franklin Paine Mall and George Corner, who urged local clinicians to avail themselves of the 'precious objects' they obtained in the course of their work. They often reciprocated by naming certain embryos after clinicians, thanking them by name in publications and passing them reports on the nature of any conditions uncovered by their research.[4]

Similar networks governed the supply of human material for tissue culture. In a 1910 paper, Alexis Carrel and Montrose Burrows outlined their belief that human tissues 'could be caused to grow in the same manner' as chick tissue, and detailed the cultivation of a tumour removed from a 35 year-old woman at the New York Memorial Hospital.[5] They outlined how the Hospital's surgeon provided the tissue in question, and thanked them for 'making possible the cultivation for the first time of a human tumor outside of the organism'.[6] The surgeon in question, a Dr. Coley, had appraised them of the operation beforehand and even allowed Carrel, Burrows or a Rockefeller technician to establish the cultures in the operating theatre – for their paper detailed how the tissue was dissected 'minutes after' its removal from the body, before being inoculated into culture medium and then transported to the Rockefeller Institute (given Carrel's insistence on rigorous technical training, we may presume this was not performed by the Hospital staff).[7]

This sample of tissue was rendered into an experimental artefact by a set of technical practices: by its dissection into fragments, inoculation in culture medium, encasement in a sealed glass apparatus, transport to the laboratory, and its subsequent maintenance and analysis in an incubator.[8] Carrel and Burrows believed these processes transformed the carcinoma sample into a 'new method for the study of human cancer'. They argued that the cultivation of human tissues would allow scientists to 'study the growth of various ... tumours and follow all morphological characters and changes of the cancerous and other cells during life', whilst also proving that human tissue could be cultivated as easily as material from other animals.[9] Yet despite these bold claims, and despite the fact that it represented the first human tissue culture, this paper did not attract any of the publicity that was later bestowed on the 'old strain'. Perhaps this was because it appeared in the *Journal of the American Medical Association* and not the *Journal of Experimental Medicine*, which the Rockefeller hierarchy promoted to journalists. A more likely explanation, however, lies in the fact that it was a relative failure. Eleven of the 12 cultures established from this sample died within five days, and the sole survivor was showing distinct signs of decay after only a week *in vitro*.[10]

Carrel and Burrows were unperturbed by this failure, attributing it to problems with the nutrient media rather than any inherent problem with human tissue. In one 1911 paper, they reported the cultivation of a breast sarcoma and, in another, detailed the cultivation of tumours from ten different cases of male and female breast cancer, as well as from a carcinoma of the lip.[11] From these samples, they established over 200 cultures. In 1914, Burrows reported how he continued to obtain and culture a 'considerable number' of tissues from the Memorial Hospital.[12] In order to determine whether pathological or normal tissue grew best *in vitro*, he acquired diseased material from the surgical removal of tumours and healthy material from operations to repair hernia. Burrows also attempted to determine whether the time that elapsed between removal from the body and cultivation *in vitro* made any difference. He established some cultures as soon as tissue was removed in operating theatres, while he carried other samples 'a considerable distance' before culturing them at the Rockefeller Institute.[13] But despite these variables, all tissues quickly died *in vitro*. Burrows observed that 'instead of active migration and growth of cells as observed about fragments of many tissues of lower animals', human tissues 'underwent a rapid dissolution' in culture.[14] Whether healthy or diseased, whether cultured in the operating theatre or the laboratory, they rarely survived more than a few days.

Yet still Rockefeller staff persisted. In 1914, Albert Ebeling and Joseph Losee predicted that once they derived a suitable medium then 'it is reasonable to suppose that human tissue can be cultivated *in vitro* for an indefinite period'.[15] Their ongoing effort to maintain human tissue reflected Carrel's desire to frame tissue culture as a potent medical tool, and the Rockefeller Institute's stated aims of making experimental biology applicable to medicine. But Rockefeller researchers were not the only ones who attempted to culture human tissue. Several researchers elsewhere were also experimenting with the types of human tissue they cultured, and with the medium they explanted it into. In 1916, the pathologist Robert A. Lambert, from Columbia University, inoculated human tissues into media obtained from different animal species and compared the growth rates of tissue obtained from living and dead bodies.[16] Lambert found that material obtained from living bodies grew marginally better than that obtained from dead bodies, and claimed that adding fowl plasma to the culture medium slightly improved the survival and growth of tissues.[17] Tellingly, he also explained why scientists persisted with human tissues, even though chick tissue appeared far easier to use. Lambert admitted he continued to use human tissue because 'it was thought that the results would be of more value if clearly applicable to human beings'.[18]

This belief also underpinned the acquisition of human tissues further afield. Across the United States, principally at Johns Hopkins University, and also in Germany, Italy, Japan and Britain, scientists utilized what the cell biologist Henry Harris called the 'good offices' of clinicians to procure human tissues.[19] In Britain, John Gordon Thomson and his brother, David Thomson, sought to culture human and chick material well before Thomas Strangeways turned his hand to tissue culture. Like Carrel, who trained David Thomson in 1913, they believed that success with human material would prove tissue culture was a 'new power' that 'will lead to discoveries of great promise'.[20] To this end, they obtained tissue from nine different operations performed at the Middlesex Hospital, London. The renowned surgeon Sir John Bland-Sutton presided over each operation; and in return for his co-operation, the Thomsons named and thanked him in all their publications.[21] As with Carrel and Burrows, it appears that Bland-Sutton allowed at least one of the brothers to wait in the operating theatre, in order to section tissue in a sterile bowl 'as soon as it was removed'.[22] Their papers document how, whilst still in the Hospital, they transferred these fragments to a test-tube and placed this in a thermos flask containing lukewarm water. They then transferred the flask and its contents to the Marcus Beck Laboratory 'as quickly as possible' and, on arrival, explanted the tissue *in vitro* and incubated it at body temperature.[23]

The Thomsons made 50 different cultures from these samples and noted that growth, if it occurred, was far slower than chick embryo tissue. Some human tissues survived long enough for them to record the proliferation of new cells from the original explant.[24] But generally, as elsewhere, the cultures did not survive long enough to provide findings of any real scientific or medical interest. By the 1920s, it was clear that human tissue could not be cultured as easily as material from other animals. At no point did Thomas Strangeways use human samples when he began tissue culture, nor does he even appear to have considered using them. Indeed, his conversion to *in vitro* methods dealt a blow to his collection of human material for pathological analysis, which he had previously acquired from Cambridge and London hospitals.[25] Strangeways relied on chick embryo tissue, and promoted it to colleagues in his *Technique of Tissue Culture*. As Edward Willmer stated, chick tissue had 'many advantages' since it was easy to obtain and grew consistently well when cultured.[26] Indeed, chick embryo material was so dependable and popular that Willmer labelled it 'standard tissue' – the very opposite of human material.[27]

These drawbacks ensured that scientists rarely sought human materials for tissue culture in Britain. Strangeways researchers only used human tissue as an adjunct to *in vitro* research, for histological comparison or morphological examination. But in some cases, they still acquired a good deal of material. For instance, the German embryologist Alfred Glücksmann, who joined the Strangeways Laboratory in 1933, obtained large numbers of biopsies as part of the ongoing radiological collaboration with St. Bartholomew's Hospital.[28] These samples were supplied by clinical contacts such as Malcolm Donaldson, the St. Bartholomew's physician and Hospital trustee.[29] In accordance with professional norms, Glücksmann and other Strangeways staff thanked their clinical suppliers in publications. Indeed, Honor Fell went so far as to name her suppliers of biopsy material in a 50th anniversary book on the Strangeways Laboratory.[30] One of these, a Dr. Mowlem, regularly stopped by to give Strangeways researchers tissue that he accumulated in the course of his work.[31] Others, meanwhile, sent biopsy samples through the post.[32] Sometimes, Fell went further than simply naming and thanking those who supplied human tissue. In 1939, she began a MRC funded project on wound healing, which sought to investigate how various factors, including age, diet and vitamin deficiencies, affected wound healing *in vitro*. This project had clear relevance to the war effort; and Fell even attempted to simulate combat conditions by growing rat tissue in culture medium that contained mustard gas.[33] She also proposed to undertake a biochemical study of the exudates from recently closed wounds, and wrote to Hadley T. Laycock, the resident surgeon at the nearby Addenbrookes Hospital, asking him to 'provide the relevant material'.[34] After Laycock agreed to obtain material by drilling holes in recently applied plaster casts, Fell named him as a project collaborator in correspondence to the MRC.[35]

Although Fell and the MRC envisaged Cambridge as the hub of important work on wound healing, the project was hampered by the wartime internment of émigré scientists on the Isle of Man, including Glücksmann and Ilse Lasnitzki, a German scientist who had recently arrived in Cambridge after spells in Italy and Denmark. Despite lobbying the MRC and the Government, Fell could not arrange their return, and the wound healing work was transferred to the MRC's Porton Down facility for the remainder of the War. Following the close of hostilities in 1945, work at the Strangeways Laboratory returned to normal, with the return of émigrés such as Glücksmann, Lasnitzki and Petar Matrinovich.

Normal service was not resumed for tissue culture in this period though. After the War, the method underwent what Honor Fell identified as a 'striking renaissance'.[36] Tissue culture increasingly moved into the medical mainstream and became a more clinically relevant, standardized and, for a time at least, a less contested technique. This 'renaissance' was reflected by the expansion of research and staff at the Strangeways laboratory. During the 1950s, the Wellcome Trust funded the construction of a large extension for biochemical research, whilst the Rockefeller Foundation paid for new equipment.[37] It was also marked by approving newspaper coverage of tissue culture research in Cambridge and elsewhere. The hopes scientists such as Frederick Spear invested in this 'infant of promise' during its early years seemed to have been finally, and widely, realized.[38]

Medical relevance and popular coverage in postwar Britain

This postwar 'renaissance' stemmed from a number of factors. For one, several technical developments ensured that tissue culture became easier. The most influential of these, the development of antibiotics, was clearly not designed with tissue culture in mind. In 1941, the British zoologist Peter Medawar used chick embryo cultures to test the safety of penicillin, which Howard Chain and Ernst Florey had recently demonstrated to have anti-microbial effects *in vivo*. Antibiotics, of course, were hailed for saving millions of lives during the War, and once they were produced in earnest, tissue cultures were regularly used to test their safety and efficacy.[39] But scientists such as Medawar also found that adding antibiotics to tissue cultures prevented the bacterial infections that regularly killed *in vitro* material. An infected culture could now be saved by a dose of antibiotics, without any detectable impact on the behaviour of tissue or cells.[40] The ability to combat bacterial infection all but removed one of the major obstacles to *in vitro* work.

At the same time, scientists began attempts to circumvent the additional problems posed by 'natural' media such as plasma, serum, and embryo extract, which often introduced unwanted variables into experiments. From the late 1940s, researchers began to develop and test synthetic media that were composed of mixtures of amino acids, carbohydrates and balanced salt solutions (although some purified biological material was often required).[41] For instance, the British-born Charity Waymouth, who worked at the Jackson Laboratory in Maine, developed different types of media that supported the rapid growth of specific tissue types for specific research purposes. These quantified media were generally more effective, more stable and easier to store than their wholly biological predecessors.[42]

These technical developments helped transform tissue cultures into more stable and productive experimental systems. Tissues, cells and organs could now be maintained for longer and utilized in research with distinct medical relevance. Nowhere was this more evident than in virology. Unlike bacteria, viruses require a host organism in which to grow. Prior to the development of tissue culture, scientists attempted to study viruses by passaging them through experimental animals. This, though, proved difficult. Laboratory animals often showed evidence of several debilitating conditions at once, which made it hard to determine exactly what effect a virus was having. At the same time, many viruses that afflicted humans, such as polio, would not grow in common experimental animals such as rats and mice. The primates that scientists consequently had to rely on were expensive to obtain and proved difficult to handle. It is no surprise, then, that virologists turned their hands to tissue culture as early as 1913.[43] But their work was generally hampered by the technical problems we have already encountered. The human tissues they used did not survive for the length of time required to adequately study viral infection.

In 1940, however, the Harvard virologist John Enders employed the 'roller tube' method of tissue culture that had been recently developed by the Johns Hopkins researcher George Otto Gey. This apparatus was an attempt to better simulate the dynamic, self-regulating environment of the body, and consisted of a rolling drum that constantly rotated test tubes full of tissue and liquid medium.[44] Rotation ensured that the liquid medium constantly 'washed' tissue, and the internal pH was maintained at a healthy level by the ongoing regulation of carbon dioxide levels. Enders found that substituting traditional culture methods for the newer roller tubes permitted a far higher rate of survival and proliferation – for both the inoculated tissue and infecting virus.[45] In a 1940 paper, Enders, T.H. Weller and A.E. Feller 'confidently' asserted that the roller tube method had distinct advantages over other ways of propagating viruses.[46] This confidence was justified during 1948 and 1949, when they used roller tubes to cultivate poliovirus in human embryonic skin and muscle for over 67 days. These tissues, notably, were grown in a medium composed mainly of balanced salts, penicillin and streptomycin.[47] As Landecker details, this work changed the field of polio research on two levels: overturning the belief that poliovirus would only grow in nervous tissue, and providing scientists with a means of viral propagation that was far easier than inoculating the virus into live primates.[48] Tissue culture subsequently became conscripted into the highly funded and publicly visible 'race to conquer polio' in the United States – lent urgency by annual outbreaks of the disease, by images of patients in iron lungs and the 1945 death of the President and polio patient Franklin D. Roosevelt.

Contemporary research on cancer also increased tissue culture's relevance to biology and medicine, by generating populations of well characterized, seemingly standardized cells that could be used in drug testing and analysis. These new cultures originated at the National Cancer Institute in Washington during the 1940s, as part of a programme designed to induce malignant changes *in vitro*. To study the processes behind malignancy, Wilton Earle and colleagues exposed cultured mouse cells to a chemical carcinogen. They found these cells underwent morphological changes associated with cancer and induced tumours when injected into animals.[49] They also, crucially, began to proliferate continuously *in vitro* and showed no sign of senescence. These perpetually dividing cultures were labelled 'strain L', and have since been classified as the first permanent 'cell line'. Cell lines offered scientists a homogenous, self-replicating and simplified experimental system. Their individual cells could be plated out, characterized and induced to multiply like bacteria, while their metabolic needs could be easily assessed.[50] Unsurprisingly, then, other scientists began to establish their own cells lines, which were soon employed across a range of medical and scientific projects. The most conspicuous example was HeLa, the first human cell line, which was established in 1951 when George Gey cultivated samples of a cervical tumour from Henrietta Lacks, a young Baltimore woman.[51] Gey soon became well known for freely distributing HeLa, ensuring that the cell line was incorporated into many national and international research projects. A 1954 article in the journal *Cancer*, for instance, noted that it had not taken long for

HeLa to become 'widely used' in cancer research.[52] HeLa also became a major tool in virological work. Between 1953 and 1955, it was grown on a massive scale in order to assay Jonas Salk's recently developed polio vaccine – overcoming a shortage of rhesus monkeys, which scientists traditionally used to test vaccine toxicity. In this short period, a specially equipped laboratory at the Tuskegee Institute produced and distributed a remarkable 600,000 HeLa cultures for vaccine evaluation alone.[53]

The vast uptake of HeLa took place amidst, and no doubt contributed to, a growing biomedical confidence in tissue culture. The skepticism of the interwar years diminished as increasing numbers of journals and textbooks extolled the method's benefits; and they were joined by new organizations such as the International Tissue Culture Association and the European Tissue Culture Club. These new organizations and medical textbooks gave practical advice on tissue culture, which they justified on the grounds that it was 'of growing importance to workers in the medical field'.[54] George Gey claimed this was due to the development of cell lines, which had freed researchers from the 'inconvenience' traditionally posed by experimental animals.[55] Biologists who used tissue cultures in drug tests also stated that it eliminated 'many of the inherent difficulties in other methods and should prove to be of value in clinical studies'.[56] This 'growing importance' was reflected by the fact that tissue culture underpinned over half of the research that was presented at the 1957 meeting of the International Society of Cell Biology, which was held in Scotland. The Society's president, the cell biologist P.J. Gaillard, claimed this offered striking evidence of the 'renewal of interest' in tissue culture since World War II.[57]

But these changing opinions were by no means confined to professional literature. Just as in the interwar years, tissue culture was the subject of considerable newspaper coverage in postwar Britain. Now, however, the tone and content of the reports differed considerably. Postwar reports on science and medicine were broadly positive: partly in a continuation of optimistic wartime reporting, which detailed how innovations such as penicillin were vital to the military effort, and partly in an effort to balance the rhetoric of postwar 'decline', which bemoaned the loss of British colonies and declining political influence *vis-à-vis* the United States.[58] Thanks to its pivotal role in the development of the polio vaccine, and its potential use in other 'magic bullets', tissue culture contributed to this celebratory climate. As Honor Fell stated in a 1960 lecture to the International Society of Cell Biology, the postwar years saw a reversal of the 'low ebb in public esteem' to which the technique had sunk in the 1920s and 1930s.[59] Whereas the press once held tissue culture as emblematic of biology's menace, reports in the 1950s were overwhelmingly positive and emphasized its medical relevance.

Announcing the award of the 1954 Nobel Prize to Enders, Weller and Robbins, a report in the *Manchester Guardian* stated that tissue culture had 'revolutionized the study of poliomyelitis' and would go on to 'form the irrevocable basis of later developments of polio vaccines'.[60] A year later, announcing the development of Salk's vaccine, the paper's American correspondent, Alistair Cooke, claimed tissue culture had 'cleared the way' for a vaccine that was the 'biggest news story of many a peacetime year'.[61] Although the jubilation surrounding Salk's vaccine was checked in 1955, when over 250 American children contracted polio from a poorly treated vaccine, this did not shake positive coverage of tissue culture. The press reported that it was central to the 'stringent safety tests' that vaccines would undergo before their introduction to Britain.[62] And in a 1957 front-page story, Cooke reported that its application in vaccine development might even have a bearing on efforts to combat cancer. Cooke detailed that when Jonas Salk cultured tissue for long periods to grow poliovirus, some cells underwent morphological changes associated with cancer, and neighbouring cells then produced antibodies that appeared to impair this malignant growth. Whilst he reminded readers that the discovery of a possible antibody against one type of cancer was a 'modest step', Cooke nevertheless predicted that tissue culture might well throw 'some light into the darkest corner of medicine'.[63] Nothing evoked clinical relevance more than a possible cure for cancer – the ultimate 'magic bullet'.

Although newspapers portrayed the polio vaccine as an American achievement, several reports detailed how tissue culture could also open 'new avenues for research' in Britain.[64] *The Times* detailed how scientists at the MRC's Common Cold Research Unit in Salisbury had used cultures of embryonic lung tissue to propagate the cold virus, and portrayed this as a pivotal step toward a vaccine for the common cold.[65] The *Guardian*, meanwhile, described how the irradiation of cultured human fibroblasts might help combat the harmful effects of atomic fallout.[66] While newspapers no longer solely associated the Strangeways laboratory with tissue culture, it continued to feature in popular coverage. Several reports detailed how Ilse Lasnitzki had applied the method to cancer research, by exposing cultured lung cells to the chemicals present in tobacco smoke and determining how this affected their structure and function.[67] The Strangeways laboratory also featured in one of the stranger reports on tissue culture's medical impact. In 1950, the *Daily Mail* and the *Daily Mirror* reported that the Strangeways researcher Margaret Hardy had managed to culture hair follicles from a mouse. Both papers represented this as a major step toward 'a cure for baldness'.[68] Hardy soon received scores of inquiries from bald men, prompting her to assert that there was 'no sign' of a cure for baldness, and that her work simply investigated how physiological factors affected hair growth.[69]

As representations of tissue culture became more positive, so did the portrayals of the material that scientists cultivated *in vitro*. Cultured cells and tissues were increasingly framed as productive factories in a microcosm. In a 1948 article for the Penguin *New Biology* series, the zoologist John E. Harris stated that cells were the site of 'most of the important transformations of matter and energy associated with life'.[70] The cell, he argued, should be regarded as a 'complex chemical workshop', powered by the cytoplasmic 'machine tools' that produced essential proteins and enzymes.[71] Similarly, 'The laboratory of the living cell' that was the subject of a *Times* report referred to both the biomedical facilities that used cells and, as the author declared, to the fact that 'a living cell is also a laboratory'.[72] During a 1962 lecture, Honor Fell also compared the cell to a 'highly efficient ICI plant operating in a space about one-sixtieth of a millimetre in diameter'.[73] Extending the analogy further, she refigured its nucleus as the 'office block' that stored and managed 'all master plans' for biochemical reactions and claimed, like Harris, that these were delegated to 'factory workmen' like mitochondria and ribosomes.[74]

According to Andrew Reynolds, the depiction of the cell as a factory reflected the fact that biochemical phenomena increasingly dominated the attention of cell biologists as the twentieth century progressed. For those interested in cellular metabolism, he argues, the factory metaphor was more resonant than portrayals of cells as elementary organisms.[75] Certainly, as Fell remarked, growing work on cell biochemistry increasingly led scientists to view cells as 'rather like an elaborate industrial concern'.[76] But the emphasis on productivity in these portrayals also suggests that they were influenced by the increasing use of cells and tissues in medically relevant work – by their role as sites of production for therapeutic 'magic bullets'. As Harris informed readers of the *New Biology*, progress in science and medicine, including the 'conquest of cancer', hinged on the ability to harness the productive capacities of cells.[77]

'An almost untapped resource': The increasing demand for human tissue

In 1951, the University of Columbia biologist Margaret Murray declared that the 'uses and advantages of cultured human tissues for medical investigations are many', and urged researchers to visit the surgical department of their local hospital to procure this 'almost untapped' resource. To Murray, operating theatres were 'unequalled as a source of human tissue'.[78] But she wrote this in the same year that George Gey established the HeLa cell line, which introduced new habits of acquisition and exchange into tissue culture. The scientists who used staggering amounts of HeLa in vaccine testing and cancer research no longer had to travel to local hospitals to get tissue, but were instead sent 'seed cultures' that they could propagate in their own laboratories.[79] During the 1950s, the circulation and storage of cell lines was aided by technical improvements in the freezing of biological material; and by the 1960s, centralized repositories such as the American Type Culture Collection began to retain and distribute large quantities of human and animal cell lines.[80]

But this did not mean that surgical theatres were displaced as a major source of human tissues. During the 1950s and 1960s, many scientists obtained human tissue in order to establish their own cell lines. As before, they acquired material from biopsies, surgical procedures or, in rare cases, from postmortems.[81] At the same time, growing criticism of cell lines ensured that scientists did not use them in a good deal of work. Researchers found that cell lines often induced tumours when injected into experimental animals, which strengthened the view that their perpetual survival *in vitro* was due to some form of malignant change.[82] This raised concern about their suitability for vaccine development, and Alistair Cooke reported how Jonas Salk would not develop vaccines on cell lines because he feared they might 'induce cancer in inoculated persons'.[83] By 1959, the fear of transmitting cancer cells prompted the US Food and Drug Administration to ban the use of cell lines in vaccine production.[84] Edward Willmer believed that all cell lines were cancerous. He detailed how cell lines derived from a variety of tissues acquired morphological characteristics of tumours *in vitro*, and questioned whether they retained any characteristics of the tissue from which they originated.[85] Willmer argued that researchers who thought they were using normal fibroblast cells may well have been working on something altogether different and 'abnormal'. He concluded that users of cell lines may not wish to 'enquire too closely about the exact nature of these cells, nor to what cells in the body they correspond'.[86]

Debates about the safety and normality of cell lines continued into the 1960s, and were often fractious. Many scientists reacted badly to claims their cell lines were innately cancerous and poorly characterized.[87] During the early 1970s, the American scientist Walter Nelson Rees argued that HeLa had contaminated many cell lines, and that researchers who thought they were working on, say, mouse epithelium were actually using human cervical carcinoma.[88] This uncertainty ensured that longstanding informal networks endured alongside the large-scale freezing, storage and transportation of cell lines. Scientists at the Strangeways Research Laboratory rarely used cell lines and, as before, they maintained good relations with surgeons at the Addenbrookes Hospital in order to ensure a ready supply of human tissue. For his work on embryonic development, Alfred Glücksmann studied the lungs of human embryos between nine and 15 weeks' development. Throughout the 1950s and 1960s, he and Victor Norfield regularly collected embryos that had been removed in surgery, after having been notified by Addenbrookes surgeons.[89] Addenbrookes doctors also provided Ilse Lasnitzki with foetal lung tissue for her work on the effects of tobacco smoke. Like Glücksmann, and as was common practice, she gratefully acknowledged their 'friendly co-operation' in her papers.[90]

The network that governed the collection of tissue remained as informal and unregulated as it had always been. Like the Thomsons in 1913, Lasnitzki and Glücksmann could simply walk to their local hospital and take whatever tissue they needed in the 1960s. This informality persisted despite the increasing regulation and criticism of medical practices and research. By the 1950s, it was clear that the 1832 Anatomy Act, which covered the use of bodies and body parts for medical education, did not adequately cover the expanding uses being made of tissue for grafting and transplantation procedures. Keen to formalize the supply of material, doctors pressed for new legislation, which led to the passage of the Corneal Grafting Act in 1952 and the Human Tissue Act in 1961.[91] Both these new Acts, however, regulated the acquisition of postmortem tissues for medical applications. They said nothing about obtaining tissues from living patients for research, which was the most common source of material for tissue culture. In 1967, the professional 'whistle-blower' Maurice Pappworth published *Human Guinea Pigs*, which criticized researchers for subjecting patients and experimental subjects to dangerous research without obtaining their full consent. Although Pappworth did criticize research involving tissue and cells, this involved tissue grafting experiments and the injection of cancerous cells into senile patients and prison inmates.[92] He said nothing about the widespread and non-consented collection of tissue samples from patients.

This lack of criticism clearly did not result from the fact that human tissues were collected and used in secret. Popular sources throughout the 1950s and 1960s openly reported that scientists used human material to develop vaccines and other 'magic bullets'. It would appear, by contrast, that this practice passed without criticism because it was widely seen as unproblematic.[93] This is important, since it undermines the claim that research on human tissues meets a deep-seated and unchanging popular resistance. Others have shown that the use of human material in research did not attract any popular criticism for much of the twentieth century. Lynn Morgan has detailed how patients in the early twentieth century did not imbue aborted or miscarried foetuses with social importance, but perceived them as waste material. This, she claims, allowed embryologists to collect them with impunity.[94] Although the acquisition of human tissue for grafting and transplantation was more high-profile and was regularly reported by American papers, Susan Lederer has likewise demonstrated that it was not controversial. Patients and journalists did not articulate concern about the retrieval of tissue, and newspapers celebrated the technical aptitude that underpinned these procedures.[95]

Similar attitudes surrounded the acquisition of human material for tissue culture in the 1920s and 1930s. Indeed, Julian Huxley ridiculed the idea of a popular attachment to tissues by framing it as one of the primitive traits held by the fictional African tribe in 'The Tissue Culture King'. Huxley detailed how the fraudulent scientist Hascombe encouraged superstitious tribespeople to maintain and worship tissue cultures he derived from the King and their deceased family members.[96] Their attachment to these 'sacred cultures' was ridiculed by Huxley's cosmopolitan narrator and stood as evidence of the 'imbecilic' mindset that allowed Hascombe to attain authority.

There remains a danger in making assumptions from this narrative alone, since Huxley was a scientist and may well have been critiquing a legitimate concern. But the fact remains that no popular source expressed concern at the acquisition and use of human materials for tissue culture. During his radio address on 'Cinematography and the Microscope', Ronald Canti described how scientists obtained tissue 'from any animal or from a human being'.[97] Canti's talk was well covered, but no-one criticized his admission that scientists sometimes obtained human material for tissue culture. When the use of tissue culture increased after World War II, newspapers were clear that scientists often cultivated human material. One *Times* report stated that Lasnitzki's work involved growing foetal tissues 'outside of the body of which they were once part'.[98] And in 1961, the *Guardian* reported how 'human embryos ending their existence at London hospitals are providing scientists with the means of making a vaccine against the common cold'. This report detailed how doctors presiding over surgical abortions notified scientists, 'who promptly collect the embryonic specimens in cold containers, tease out the still-living cells, and dispatch them to the MRC's cold research unit at Salisbury'.[99] Even when the use of foetal tissue cultures was outlined in depth, these newspapers voiced no criticism. In line with the broadly positive climate, which celebrated medical and scientific innovation, they portrayed the cultivation of human tissues as a 'technical advance of first-rate importance'.[100]

As in earlier chapters, we thus see homology between scientific practices and popular representations of human tissues. In surgical theatres, scientific publications and newspaper reports, they were comprehensively treated as raw materials of biomedical research. The issues of consent and property we encounter today were not concerns for most of the twentieth century – for scientists or the journalists who covered their work. Analysing how and why they were 'overlooked' therefore leads us down a blind alley. We should instead seek to explain why they have come to prominence in recent decades.

Conclusions

Ethicists sometimes frame the informal collection and use of human tissue as a 'sad commentary' on how biomedical practices contravened popular values for much of the twentieth century.[101] Yet this thesis erects an untenable barrier between material practices and their wider culture – overlooking how the exchange of tissue helped constitute a broader fabric of social relations. As Catherine Waldby and Robert Mitchell state, the processes of collection, storage and distribution that human tissues undergo are determined by the particular significance they have for different actors at different times and places: for clinicians, patients, scientists, the media and so on.[102] Since the collection and use of human tissue was clearly outlined by newspapers, and since it attracted no criticism, it would appear that the views of these various parties did not come into conflict over the period detailed here.

Bioethical and popular accounts that flag the non-consented acquisition of tissues as evidence of professional malpractice rather miss the point, then. Uprooting contemporary ethical arguments, like getting patient consent for research on tissues, and applying them to periods where they were not considered obfuscates our appreciation of both the past and the present. It presumes that attitudes to tissue remain constant and overlooks the interplay that endows them with meaning in both science and society. And it weakens our appreciation of how ethical principles like informed consent are not timeless, but are rather the product of historically specific social, political and professional factors.

5
'A Cell is Not an Animal': Negotiating Species Boundaries in the 1960s and 1970s

During the 1960s and 1970s, tissue culture featured prominently in two debates about species differences. The first stemmed from the work of Henry Harris and John Watkins, who successfully fused human and mouse cells to create cross-species hybrid cells in 1965. To journalists, popular writers and documentary makers, these hybrid cells heralded a new era of control over nature – a 'biological revolution'. But as before, many of the most sensational claims regarding hybrid cells came from their creators. Harris and Watkins deliberately fused human and mouse cells in order to draw attention to their work, and publicly extrapolated from hybrid cells to chimaeric organisms such as human-ape 'mapes'. Their material practices and populist rhetoric drew upon popular fascination with species boundaries, in order to resonate with the contemporary 'biological revolution'.

However, the transgression of human species barriers also evokes pollution and transgression, suggesting violation of the order and hierarchy central to human identity.[1] Since the first appearance of the fire-breathing chimaera in the Greek myths, the conflation of human and animal has evoked hellish disorder, Divine wrath and impending disaster.[2] It should come as no surprise, then, to find that hybrid cells attracted considerable criticism in the 1960s. But as Lorraine Daston and Catherine Park note, we must be attuned to 'the audiences, historical meanings and circumstances that shape and nourish' responses to species transgression.[3] In this case, negative representations of Harris and Watkins's cells reflected a post atomic mindset that was increasingly aware of science's potential risks, and sceptical toward traditional seats of authority. Rather than stand for a biological revolution, as Harris and Watkins intended, hybrid cells instead embodied what one writer pointedly called *The Biological Time Bomb*. To dampen the resultant criticism of their work, Harris and Watkins publicly reasserted the boundaries they had claimed to efface – between cells, animals and species.

The second debate involved claims that human tissue cultures offered superior alternatives to experimental animals, by ensuring that one did not have to extrapolate across species barriers to ascertain what effect a drug may have in patients. This claim was first articulated by welfare-minded scientists in the 1950s and was taken up by a resurgent anti-vivisection movement in the late 1960s. During the early 1970s, it found favour with politicians and the press. But scientists like Honor Fell, now in her seventies, responded by pointing to the fundamental difference between tissue cultures and animals. She argued that because cultured cells lacked physiological, circulatory and nervous systems, they were poorer substitutes for the human body than animals of a different species.

Creating hybrid cells

Buoyed by tissue culture's increasing use in virology and drug testing, scientists in other fields also began to employ the method during the early 1960s. One of the most conspicuous examples was in cell genetics. Here, as the cell biologist Henry Harris attested, progress in studying the genetics of non-sexual, or 'somatic', cells from higher animals had been curtailed by the lack of a method that allowed them to be individually studied and induced to fuse. Experimental cell fusion was common practice in the genetic study of bacterial and fungal cells. It provided hybrid progeny with quantifiable genetic 'markers' from their respective parent cells; and analysing how these markers combined and interacted in the hybrids permitted investigation of gene expression and regulation. Until the 1960s, scientists lacked a practice whereby the non-sexual cells of higher animals could be studied in this way, and appreciation of somatic cell genetics lagged behind that of bacterial and fungal cells.[4]

The development of cell lines during the 1950s and 1960s offered a solution to this impasse. Their constituent cells could be individually plated, genetically studied and cloned in the same fashion as bacteria or fungi. Additionally, in 1953 John Enders discovered that the addition of viral particles to tissue cultures often caused cells to fuse.[5] Giant cells containing more than one nucleus had been described in pathological samples since the nineteenth century, and Thomas Strangeways had reported their presence *in vitro* during the 1920s, but Enders was the first to suggest that a virus caused cells to fuse. By 1958, other virologists observed that the mumps, influenza and infantile croup viruses also induced cell fusion. Drawing on these observations in 1962, a group of Japanese biologists led by Yoshio Okada treated a haemagglutinating virus with ultra-violet radiation, before adding it to a mouse cell line. As hoped, they found that the weakened virus fused cells together.[6]

Henry Harris undertook his own research on cell fusion when he left the John Innes Institute, in Norwich, to take the chair of pathology at the University of Oxford in 1964. Harris was primarily interested in assessing the genetic differences between cancerous and normal cells, and believed that cell fusion provided a good number of genetic markers to investigate and measure the action of certain genes in cancer.[7] Working with the Oxford virologist John Watkins, he began by selecting which cells to fuse. Harris and Watkins notably altered Okada's protocol in one respect and elected to fuse cells from two different species. On the face of it, concrete practical reasons lay behind this decision. Harris and Watkins believed that crossing cells from different species would provide a greater supply of genetic markers.[8] The cells to be fused appeared to have been chosen for similarly practical reasons. The human HeLa cell line and mouse Erlich ascites cells could both be obtained in large quantities, could be individually isolated and were known to survive well *in vitro*. Importantly, both also had morphologically distinct nuclei that could be easily distinguished in any cell hybrid.

On 25 October 1964, Harris and Watkins mixed HeLa and Erlich ascites cells in a test tube containing an inactivated strain of the Sendai virus, then incubated the mixture overnight. Before the experiment, they had labelled the HeLa nuclei with a radioactive marker and left the Erlich ascites cells unlabelled. The following day, Harris and Watkins observed large cells that contained both labelled and unlabelled nuclei: offering direct evidence that cells from different species had been artificially fused for the first time. In the paper that announced this feat, published in *Nature* on 13 February 1965, Harris and Watkins claimed their hybrid cells survived *in vitro* for up to 15 days, moved 'sluggishly' and even synthesized RNA, protein and DNA.[9] This latter property, they argued, showed that 'genes of mouse and man are therefore transcribed' in the hybrids. What is more, the human and mouse nuclei appeared to fuse in some cells, and these mononuclear hybrids went on to divide. This, Harris and Watkins argued, permitted the study of gene expression and regulation in a fashion that had only previously been possible in bacterial or fungal cells. They ended their paper by confidently stating that cross-species hybrid cells facilitated modes of analysis 'which have hitherto not been possible in animal cells'.[10]

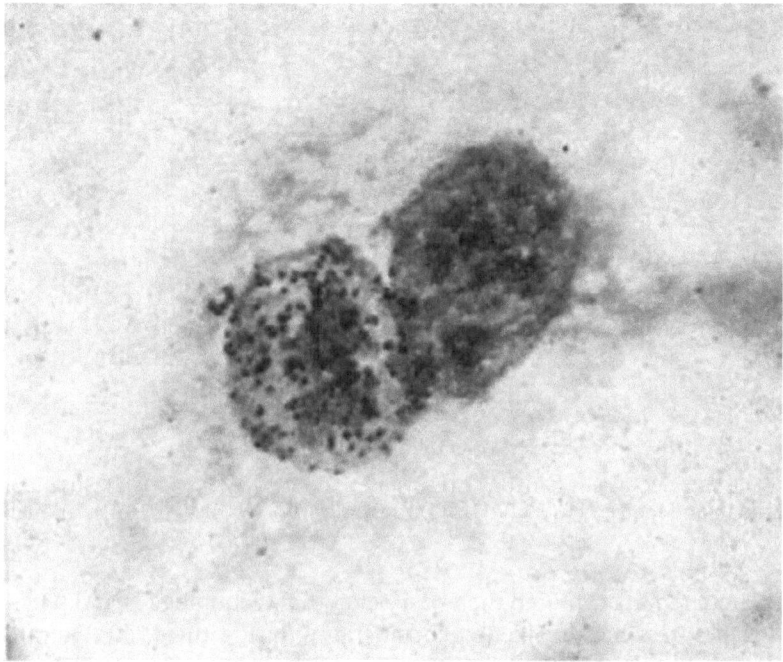

Figure 5.1 Autoradiograph of a hybrid cell containing one dotted, i.e., radio-labelled, HeLa nucleus and the unlabelled Erlich ascites nucleus.[11]

The press immediately seized upon the apparent demolition of species barriers. On the same day as Harris and Watkins published their paper in *Nature*, *The Times* stated they had produced 'living cells that are part mouse and part man'.[12] The following day, a front-page story in *The Sunday Times* described the cells as the 'strangest hybrid form of life ever seen in the lab – or out of it'. These hybrids, it continued, had survived for up to two weeks and 'may actually reproduce'.[13] On 15 February, the *Daily Mirror* launched what was to become a dominant reaction to cell fusion, with an illustration that extrapolated from hybrid cells to imagined chimaeras (see Figure 5.2). Within weeks, the *New York Times* and newspapers throughout Europe had also published stories on the hybrid cells.[14]

"*Who was Walt Disney, Dad?*"

Figure 5.2   From cells to hybrid organisms: the *Daily Mirror* reports the fusion of human and mouse cells in 1965.[15] Courtesy of Mirror Syndication International.

These reports captured the wider belief that technological and scientific progress had quickened considerably during the 1960s. Harold Wilson's Labour Party won the 1964 general election on a promise to turn what he called the 'staggering rate' of innovation into material and economic prosperity.[16] At the same time, new periodicals like the *New Scientist* and television programmes like *Tomorrow's World* reported the frontiers being broken by science and medicine.[17] There was an increasing sense, in particular, that biologists were undertaking the most influential work. In his 1959 book *Can Man Be Modified?*, the French writer Jean Rostand argued that work on artificial fertilization, the induction of genetic mutations, the nature of DNA and the development of psychotropic drugs showed that biology 'has reached a point of maturity at which its discoveries and inventions have more or less powerful repercussions upon the fate of individuals or the species'.[18] Newspaper coverage of this work claimed the 1960s heralded a 'biological revolution', with biologists attaining an unprecedented ability to control and redefine the natural world.[19] This sense of power stemmed in part from the confident rhetoric of the young biologists who featured in popular coverage. In 1967, the American biologist Robert Sinsheimer, who had recently synthesized DNA *in vitro*, proclaimed that:

> Dramatic advances have laid bare the essential mechanisms of inheritance and of the processes of cellular function and control ... As we understand life we can control life. This has been a historic pattern in the physical sciences and we have today a vast control of our physical environment. We will soon be acquiring a similar control of the biological world; and now this impact will strike home, for the biological world includes us. In all of science we have been, in a sense, children. But we in science are growing up now; our toys become more potent. The games we play with nature have great stakes and their outcome moves the whole social structure.[20]

Like Sinsheimer, Henry Harris also saw his work as revolutionary. As he admitted in a later interview, he and Watkins could have joined the cells of any two animals, but deliberately chose cells from 'two species long regarded as antithesis' in order to secure public attention.[21] Indeed, the fusion of human and mouse cells was as much a public demonstration of biology's power as a simply practical choice. This is reinforced by the fact that Harris publicly dwelt on the sheer novelty of his and Watkins's hybrid cells. Writing in the *New Scientist* in February 1965, he stated that 'none of the information which will flow from these cells will be quite as exciting as the initial discovery that they could be produced at all'.[22] And in a long *Tit-Bits* report that portrayed hybrid cells as evidence of the 'fantastic power' biologists now yielded, Harris claimed that:

> The production of artificial hybrids is interesting. It may possibly prove to be a boon to farmers. And to others, in time. But our own interest lies in trying to cross the frontiers of nature. To show that it can be done, and then to see what use it can be put to ... We are creating something entirely new. New life. These cells of ours are forms of life which have never existed before ... We can take cells which Nature refuses to join together and simply make a cross between them.[23]

Like the *Daily Mirror*, Harris also extrapolated from hybrid cells to chimaeric organisms. In *Tit-Bits*, he hinted that chimaeras were feasible by stating that 'most would certainly be infertile, like mules. But many of them could be very useful. I don't believe it's beyond the wit of man to design a use for them'.[24] And in an episode of the BBC documentary *Towards Tomorrow*, he argued that the cells of mice and rats could be fused to form a new organism, called a 'mat', and that the fused cells of humans and apes could generate a 'mape'.[25]

Hybrid cells and *The Biological Time-Bomb*

Although many commentators agreed that the potential of science and technology had increased in the 1960s, they presented these possibilities in a more ambivalent light than the generally positive reports from the 1950s. This was encapsulated by the Cambridge anthropologist Edmund Leach's 1967 BBC Reith lectures, entitled *A Runaway World*. 'Science increasingly offers us total mastery over our environment and our destiny', Leach argued, 'yet instead of rejoicing we feel deeply afraid'.[26] The physicist Eric Burhop agreed that the 1960s had witnessed 'the growth of a disenchantment with our modern technological society', despite postwar advances such as 'transistors, computers, nuclear power, insecticides and antibiotics'.[27] As Burhop conceded, this disenchantment had several roots.[28] For one, from the late 1950s new social movements such as the Campaign for Nuclear Disarmament (hereafter CND), and prominent members including Bertrand Russell, increasingly drew attention to the destructive capabilities of thermonuclear weapons. In his 1961 book *Has Man a Future?*, Russell claimed that the existence of nuclear weapons raised a simple, but profound, question. 'Is it possible', he asked, 'for a scientific society to exist, or must such a society inevitably bring itself to destruction?'[29]

This, of course, is similar to criticism of science found during the 1930s. But debates now had greater immediacy and a higher profile; thanks to the hitherto unimaginable scale on which dangers were now conceived, and to the fact that criticism emanated from a broader social base than before. The unease voiced by Russell and CND also came from an activist environmental movement that rallied around Rachel Carson's 1962 book *Silent Spring*, from an increasingly critical media and from intellectual critiques of modern science and technology like Jacques Ellul's *The Technological Society*. At the same time, representations of medicine as the source of untold 'magic bullets' all but vanished following the 1962 disclosure that the morning sickness pill, Thalidomide, caused significant birth defects. As Leach detailed, disenchantment with science was particularly strong amongst an increasingly radical young generation, who harboured a deep suspicion toward traditional seats of authority and saw science's role in the production of chemical weapons as evidence of its subversion to political interests.[30] But while members of the new counterculture may have framed themselves as its torchbearers, this crisis of confidence was by no means their preserve alone.[31] Ambivalence was voiced across the political spectrum, in tabloid and broadsheet newspapers, and by young and old alike.

By the late 1960s, the 'biological revolution' was attracting the most concern. Once held as the major route to catastrophe, atomic and thermonuclear weapons now acted as a guide against which the threat from biology could be gauged. As the biologist Steven Rose and his sociologist wife Hillary Rose argued in 1969, there was a sense that 'the problems raised by nuclear power, though they are still unsolved, may be drastically overshadowed by the equally dramatic technological changes of ... the biological revolution'.[32] In a book titled *The Biological Time-Bomb*, the journalist Gordon Rattray Taylor claimed the possible outcomes of biological research, just as in nuclear weapons, outweighed any conceivable advantage. Taylor warned that if left unaddressed, the issues raised by the growing interference with heredity, mental behaviour, reproduction and life span threatened 'nothing less than the break-up of civilization as we know it'.[33]

The Biological Time-Bomb was an immensely pessimistic book. Its dedication was signed 'with love and foreboding', whilst the final chapter was titled 'the future, if any'.[34] Other popular sources were similarly foreboding and echoed Taylor's assertion that 'the world is bent on going to hell in a handcart'.[35] Reviewing the book for *The Times*, the playwright Dennis Potter outlined how it had become 'the taste of the times to look around the laboratory, then to look ahead and shudder'.[36] To Potter, this sense of foreboding was only natural, since 'If the research biologist can tamper with the genetic structure of human beings, then there seems to be no place left to retreat, no last citadel which cannot be stormed, no secret little hope or delusion that will not be betrayed'.[37]

The conflation of species boundaries was often held as evidence of the dangers posed by the 'biological revolution'. Gordon Rattray Taylor claimed that interference with heredity and species boundaries raised 'the most serious of all the human problems created by biological research'.[38] The fact that hybrid cells survived 'and still function' *in vitro*, he argued, made it likely that cell fusion would soon be combined with genetic engineering as part of a 'new eugenics'.[39] Taylor predicted that cell fusion would allow scientists to transfer DNA between cells, introduce foreign genes and create new organisms. A likely outcome of this, he concluded, was the implantation of human nuclei into animals, 'perhaps apes', in order to produce hybrid organisms that would function as menial labour.[40] Taylor went so far as to outline the legal dilemmas these chimaeras may raise. What, he wondered, 'will be the legal status of a creature with human chromosomes, but animal appearance? And conversely?' Furthermore, he questioned how like a human a chimaera would have to be to 'qualify for human benefits, including access to the retirement pension or to union membership?'[41]

Although Taylor displayed the transgression of species barriers as an unwelcome prospect, the writer Arthur Koestler claimed that predictions of chimaeras marked the point where 'the shudders give way to the giggles'. Whilst Koestler did not dispute that cell fusion made chimaeras likely, he questioned whether they would be greeted with horror and made to serve strictly utilitarian ends. He argued that if certain biologists possessed an artistic temperament, then 'there is hope that 50 years from now there will be centaurs capering through Kensington Gardens, and mermaids offering cups of tea to sailors on the Serpentine, while goaty Pan will blow his pipe under the Albert Memorial in an aerosol cloud of deodorant'.[42] Generally though, popular coverage echoed Taylor's pessimism. A *Tit-Bits* report claimed that cell fusion offered chilling evidence of how 'biologists are taking over as Big Brother from where the physicists left off'.[43] In a report entitled 'Why Scientists Create Monsters', *Tit-Bits* used Harris's claim that 'our interest lies in trying to cross the frontiers of nature' to argue that cell fusion was simply a demonstration of biological prowess, which served no practical ends. Scientists created these 'monsters', in other words, simply because they could. 'What', it pointedly asked, 'was the purpose of such an experiment?'[44]

Similar questions were raised by a BBC documentary on the 'biological revolution'. This was part of the *Towards Tomorrow* series, which reflected the futuristic tenor of popular culture during the 1960s, as well as the increasingly critical tone of television reporting (the opening credits set scenes of industrial pollution and dying animals to the Hedgehoppers Anonymous song 'Good News Week').[45] Each week, *Towards Tomorrow* surveyed scientific research to pose the question that 'your future is being created now – for better or for worse?'[46] Some weeks, it was answered positively. Robots, for instance, were portrayed as technologies that 'may bring about salvation for the human race'.[47] Biology, however, was a different matter. The episode that looked at work such as cell fusion, *in vitro* fertilization and genetic engineering carried the foreboding title 'Assault on Life'. The episode's producer chose hybrid cells as the embodiment of this 'assault', stating in the *Radio Times* that:

A mat is part mouse, part rat. What's a man-ape? A 'mape'? In Oxford a scientist already fuses living human and mouse cells ... What for?[48]

When it was screened, the episode focussed on the grave 'legal, social, religious, philosophical and spiritual' questions that biology raised.[49] Harris was filmed at work and, on camera, justified the hypothetical production of 'mats' and 'mapes'. Calling for strict regulation of such research, the episode argued that scientists such as Harris could no longer be trusted to work on what they chose, without considering the impact on society. Nuclear weapons were again evoked to justify this unease, with the narrator ominously stating that 'Rutherford also said splitting the atom would serve no practical purpose'.[50]

Within days of the episode's transmission, newspapers carried letters voicing anxiety at cell fusion. One correspondent to *The Times* stated how she watched with 'increasing horror and fear ... while scientists try and create different forms of life by combining the cells of different species, i.e., man and mouse, man and monkey, &c'.[51] Cell fusion, she continued, 'would be disturbing science fiction, were it not for real'. Echoing the programme's conclusion, this correspondent argued that work on cell fusion should be suspended immediately, since 'the atom bomb is enough evidence to show what society will do with a scientific discovery'.[52] Several respondents to a BBC viewer survey agreed, considering Harris and Watkins's experiment to be 'nothing more than men trying to be Gods'.[53]

Over the following days, Harris and Watkins attempted to counter the ambivalent portrayal of cell fusion. Writing in *The Times*, Harris complained that their work had been 'grossly misrepresented' by *Towards Tomorrow*.[54] Harris now dwelt on the potential uses of cell fusion, asserting that the hybrid cells offered a route to understanding how genes were controlled. He claimed this would provide important 'new information about cancer', where the genes that promote cell division were 'permanently switched on'.[55] He and Watkins also asserted the fundamental differences between hybrid cells and chimaeric organisms. Harris stressed in *The Times* that fusing cells together 'is not the same as creating a new species. *A cell is not an animal*'.[56] In a long letter the following day, Watkins also asserted that 'it is cells of different animals that are being fused together, and not whole animals'. A cultured cell, he claimed, bore 'about the same relationship to an entire organism as a single brick bears to a city'.[57] By arguing that hybrid cells were not animals, Harris and Watkins attempted to place the blame on media ignorance and misreporting. Harris claimed that popular anxiety surrounding hybrid cells resulted from the media's 'ability to distort, misrepresent and terrify'.[58] Watkins, meanwhile, argued that controversy was inevitable, since 'those responsible [for *Towards Tomorrow*] did not allow viewers to see the purposes behind experiments'.[59]

But this argument falls short when we consider that Harris and Watkins sensationalized their hybrid cells from the outset. They deliberately chose to combine human and mouse cells because the two species had long been popularly viewed as antithetical. In *Tit-Bits* and the *New Scientist*, Harris claimed their main objective lay 'in trying to cross the frontiers of nature'. This is to say nothing of his public predictions of rat-mice 'mats' and man-ape 'mapes'. Like 'chemical babies' in the 1930s, popular representations of human-animal chimaeras in the 1960s did not arise from a misinformed public mindset. They arose from, and were heavily justified by, scientific attempts to promote cell fusion.

Fortunately for Harris and Watkins, controversy surrounding hybrid cells abated as the 1960s came to an end. This was not due to popular acceptance of species effacement, since the transplantation of animal parts into humans continued to raise disquieting ethical, legal and social questions into the 1970s and 1980s.[60] Nor was it due to growing support for biological research, since many experimental practices remained contentious. Instead, controversial work in other fields displaced hybrid cells as troubling emblems of the 'biological revolution'. Even when cell fusion was controversial, newspapers and popular books consistently held the *in vitro* development of human babies to be the major outcome of the 'biological revolution'.[61] Perhaps unsurprisingly, then, the 'test-tube baby' became the recurring emblem in popular debates after February 1969, when the Cambridge physiologists Robert Edwards and Barry Bavister, and the Oldham obstetrician Patrick Steptoe, announced the fertilization and maturation of a human oocyte *in vitro*.[62] But though the motif changed, the nature of discussion remained largely the same. Like 'mapes' before them, newspapers framed test-tube babies as evidence of biology's underlying dangers. Scientific responses were homologous too. Like Harris and Watkins, Robert Edwards publicly asserted the medical implications of *in vitro* fertilization, and differentiated between cultured embryos and mass-produced human clones.[63]

So too, notably, did Honor Fell. Although Robert Edwards was now the centre of media attention, the Strangeways Research Laboratory was still implicated in some predictions of 'test-tube babies'. Gordon Rattray Taylor, for one, claimed that the laboratory's work on embryo culture offered evidence that the mass-production of human babies 'does not seem unduly difficult'.[64] In the outline of tissue culture's 'past, present and future' that she gave in a 1969 lecture, Fell dwelt on how the Strangeways Laboratory had 'been intermittently plagued by the "test tube baby"'. She detailed how the popular press were once again 'convinced that the ultimate aim of laboratories like ours is to develop a method for growing humans in culture from fertilization up to the stage of birth, so that the population can be mass-produced in factories'.[65] To Fell, this scenario marked the 'non-future' of tissue culture, as 'research on mammalian embryos in culture is undertaken with the sole purpose of investigating developmental processes, and *not* with the object of establishing a human baby-factory'.[66] Fell claimed that even the most sophisticated culture vessel was 'childishly simple, crude and ineffectual' in comparison to the human womb, and assured her audience that the 'test tube baby is no nearer realization now than it was 40 years ago, and I confidently predict it will be equally remote 40 years hence'.[67]

But when covering the 'popular ideas now current' about tissue culture, Fell could have mentioned the growing body of opinion that framed human tissue cultures as replacements for experimental animals. As the *Guardian* reported in 1968, anti-vivisection groups had recently begun to argue that tissue culture relieved the brutality of animal experimentation and, crucially, overcame the extrapolation across species barriers that occurred when scientists used animal models to predict the action a drug would have in humans.[68] They claimed that tissue culture was 'less haphazard' than vivisection and called for the replacement of experimental animals with *in vitro* methods.[69] During the 1970s, as 'animal rights' became more prominent, and vivisection more contested, this argument was increasingly supported across the press and in Parliament. To Honor Fell, however, it was as misleading a portrayal of tissue culture as predictions of 'test-tube babies'.

Replacing experimental animals with tissue culture

In his 1930 article for the *Sunday Express*, Oliver Lodge had claimed that tissue culture offered a better means of experiment than 'watching the comfort or discomfort of live animals subjected to treatment'.[70] Whilst few argued that vivisection was a useless procedure by the 1930s, Lodge outlined how some prominent figures, like his friend George

Bernard Shaw, continued to 'doubt whether the information gained is sufficient to justify the subjection of living creatures to pain and discomfort'. Lodge argued that tissue culture might provide a solution to this dilemma, by allowing experiments to be conducted 'under circumstances in which observations can be made in the most direct manner, without pain or discomfort of any kind'. With this in mind, he voiced his hope that:

> the activities of kindly people, which have hitherto been exercised in a negative direction, will now take a positive turn, and that they will ease their consciences and use their resources to encourage the new methods of investigation and to promote the study of vital activities and cellular changes in the form of microscopic tissue-cultures wherever possible.[71]

Researchers at the Strangeways Research Laboratory had sometimes framed cultured cells as experimental animals. In a 1928 lecture, for instance, Frederick Spear detailed how tissue culture refigured 'The Cell as an Experimental Animal'. And in her 1930 radio talk, Honor Fell outlined how the method allowed scientists to study 'living cells as easily as living tadpoles'.[72] Writing in the *British Journal of Radiology* in 1935, she declared tissue culture's major advantage was that it:

> provides us with a method of investigating the reactions and potentialities of cells, and small fragments of tissue when removed from the general influence of the body, *i.e.*, when deprived of a vascular system, nerve supply, and association with normally adjacent parts. In other words, we are able to use living cells as experimental animals.[73]

In the same article, however, Fell claimed that tissue culture could never replace experimental animals, since 'we seldom expect to obtain the same results *in vitro* that we get *in vivo*'.[74] Whilst tissue culture facilitated more precise analysis of the effect chemicals had on cells, Fell believed that it was a poor substitute for the body due to its elimination of numerous 'complicating' factors, such as nervous and circulatory systems. She concluded that 'one of the greatest advantages of the tissue culture method, *viz.*, the simplification of experimental conditions which it implies, is at the same time one of its greatest limitations'. Scientists should therefore bear in mind, she cautioned, that 'work *in vitro* can often give us only one chapter of a story that may run into several volumes'. They would still have to rely on experimental animals to obtain information on 'the general physiological processes of the body as whole', and should only regard tissue culture 'as a valid and enormously important accessory technique to work done *in vivo*'.[75]

But Lodge's article aside, there is no evidence to suggest that tissue culture was seriously considered as an alternative to experimental animals during the 1920s and 1930s. No other newspaper carried a similar argument, and no anti-vivisection or animal welfare organization promoted tissue culture. By the 1930s, animal experiments were considered to be unfortunate but necessary procedures. The calls for an outright ban and the sizeable public support that marked Victorian anti-vivisection campaigns all but vanished in the face of new therapeutic agents that were developed from the turn of the century. These offered persuasive evidence of vivisection's medical relevance; and as research expanded in pharmacology, bacteriology and physiology, the British government issued greater numbers of licences for animal experiments.[76] With agreement on the need for vivisection held by all but a committed abolitionist fringe, attention turned to the conditions in which experimental animals lived. Newer organizations such as University of London Animal Welfare Society, founded in 1926, promoted good laboratory conditions and animal husbandry; but these were mainly scientific groups and spoke predominantly to professional concerns (in 1938, the society changed its name to the Universities Federation for Animal Welfare, hereafter UFAW).[77] At the same time, tissue culture was viewed as anything but a reliable alternative. Journals such as the *Lancet* openly questioned its medical relevance, while researchers struggled to maintain human tissues *in vitro*.

Animal welfare groups did not seriously consider tissue culture as an alternative until the late 1950s, when the method was increasingly used across a range of disciplines. During a 1957 conference on 'Humane Techniques in the Laboratory', Kingsley Sanders, the director of the MRC's viral research group, spoke about the 'revolution which modern techniques of tissue culture have started in virology'.[78] Sanders claimed that the application of antibiotics and the development of standard media had removed the 'aura of difficulty' from tissue culture, and made it suitable for 'any reasonably equipped laboratory'. Indeed, he argued that provided with some tissue, an incubator and a supply of utensils 'even an amateur in his kitchen can do it'.[79] Sanders framed the development of the polio vaccine as evidence of tissue culture's newfound potential, outlining how the vaccine would not have existed were it not for work on human tissue.[80] He stated that tissue cultures were superior to experimental animals because they were cheaper to obtain, easier to handle and gave a greater yield of virus. Although he conceded that animals may have to sometimes be killed to establish a tissue culture, Sanders stressed that the use of human tissues would 'supplant the experimental animal entirely'.[81]

Sanders did not consider any humanitarian implications of the shift to tissue culture. He simply focussed on the benefits to scientists, who were able to handle viruses with greater precision and were 'liberated from the hazards and uncertainties of animal experiment'.[82] But another paper at the conference did outline how tissue culture might benefit experimental animals. Here, the zoologist William Russell, who UFAW employed to investigate humane laboratory techniques, set out what he called the 'Three Rs' approach: involving reduction in the number of animals used in experiments, refinement of experimental procedures and replacement of animals with alternative methods.[83] In this paper, and the 1959 book *The Principles of Humane Technique*, co-written with Rex Burch, Russell outlined the merits of several potential replacements, including microorganisms, computer models and tissue culture. Of these, tissue culture appeared to be the most promising. Russell and Burch stated that 'mammalian tissue cultures have become, since the Second World War, one of the most important replacing techniques, and indeed one of the most important developments in biology'.[84] Like Sanders, they predicted that the use of human tissues promised to eliminate the experimental animal altogether – transforming tissue culture into 'the absolute ideal' for animal welfare campaigners'.[85]

Russell and Burch also presented tissue cultures as better models of the human body than animals of a different species. They sought to dismantle the 'fallacy' that stated all bodily properties, such as physiological and circulatory systems, had to be present in an experimental model, arguing that the human body was a 'reducible system' whose functions could be separately studied without recourse to vivisection.[86] They claimed an experimental system that reproduced only one of these component functions – i.e., tissue culture – often generated more accurate results than an animal which had been forced to act as proxy for the whole system. To support this claim, Russell and Burch documented tests for the antibiotic cycloserine, which was found to be inactive in experimental mice but highly potent in cultures of human skin and, crucially, in patients.[87] This, they argued, highlighted how 'the fidelity of mammals as models of man may be greatly overestimated'.[88]

But despite Russell and Burch's efforts, tissue culture was largely overlooked as an alternative to experimental animals for much of the 1960s. A Home Office inquiry into vivisection, chaired by the lawyer Sidney

Littlewood and completed in 1965, recorded that animal experiment-ation had actually increased since the late 1950s. Littlewood's report ascribed this to a heightened concern for patient safety after the Thalido-mide scandal, which ensured that scientists and companies now sub-jected drugs to more rigorous animal tests.[89] Yet it also argued that animal experiments continued to rise because no method currently offered a viable alternative. Following extensive consultation with scientists, Little-wood concluded, in contrast to Russell and Burch, that 'discoveries of adequate substitutes for animal tests have so far been uncommon and we have not been encouraged to believe that they are likely to be more frequent in the future'.[90]

Littlewood's report also claimed that the general public viewed animal experiments as a 'disagreeable necessity', with protest emanating from a committed but marginal band of anti-vivisectionists.[91] But by the close of the decade, and throughout the 1970s, animal experimentation became increasingly contested. Publications such as the *Guardian,* the *Daily Mirror*, and even the *New Scientist* began to question the widespread use of animals for research and teaching. Investigative reporters for the *Sunday People* condemned the use of dogs in smoking research, and increasingly militant anti-vivisectionists broke into laboratories to 'lib-erate' experimental animals. Letters to the Home Office protesting against vivisection increased ten-fold: from 230 in 1972 to over 2,500 in 1975.[92] And members of the House of Commons and the House of Lords attempted to introduce more restrictive legislation, claiming that animal experiment-ation had become a 'matter of the greatest concern to a large number of people'.[93]

This criticism reflected a number of factors. Anti-vivisectionists began to align themselves with the environmental movement and high-profile campaigns for racial and sexual equality. Books such as John Vyvyan's *In Pity and In Anger*, published in 1969, argued that animal experimenta-tion was another example of 'the pitiless exploitation of the rest of nature for the physical benefit of man'.[94] At the same time, a group of Oxford academics called for 'a fresh look at the whole subject of experiments on animals'.[95] Throughout the 1970s, the psychologist Richard Ryder, a member of this Oxford group, drew upon civil rights campaigns to des-cribe animal experiments as a form of exploitation he labelled 'speciesism'. Ryder argued that like sexism and racism, exploitation of animals was justified by recourse to superficial physical differences.[96] The same year, the philosopher Peter Singer, who had studied at Oxford during the late 1960s, published his influential *Animal Liberation*, which claimed that humans had no natural or moral right of dominion over animals.

Experimenting on vegetative humans, Singer argued, offered a morally acceptable alternative.[97]

Animal experimentation also came under fire thanks to the increasing presentation of tissue culture as a failsafe alternative. This campaign gave the anti-vivisection movement greater credibility into the 1970s.[98] Anti-vivisectionists could now point to the existence of alternative methods when they wished to frame animal experiments as unnecessary; and their support for certain laboratory techniques meant they could not be simply dismissed as sentimental and anti-scientific. As the *Guardian* noted in 1972, support for tissue culture had transformed the anti-vivisection movement from 'cranks to an effective pressure group'.[99] The prime advocate of this new tactic was the welfare campaigner Dorothy Hegarty, who called for anti-vivisectionists to adopt a 'new, positive approach' by promoting alternatives.[100] Hegarty urged groups such as the British Union for the Abolition of Vivisection (hereafter BUAV) to 'put all you have into promoting the idea of using substitutes for animals, because this is the surest and quickest way of relieving their suffering'.[101] The following year, having secured patronage from the wealthy Countess of Kinnoull, Hegarty and the retired biologist Charles Foister established an organization to do precisely this. Run from a small office in Wimbledon, south London, the Fund for the Replacement of Animals in Medical Experiments (hereafter FRAME) publicized alternative methods by writing to scientists, newspapers, biomedical journals and every member of both Houses of Parliament.

Though FRAME essentially reiterated the Three Rs approach, the heightened tension over animal experiments in the late 1960s and 1970s granted them a higher public profile than Russell and Burch. FRAME's early campaigns were covered favourably by Bernard Dixon, editor of the *New Scientist*, who endorsed the promotion of alternatives over the absolutist arguments of 'the outright vivisectionist'.[102] Foister and Hegarty also seized upon on particular controversies to publicize alternatives. Following the outcry over 'smoking dogs' in 1971, Foister wrote to *The Times* demanding 'practical alternatives' to experimental animals.[103] To illustrate that such alternatives existed, and to highlight the inhumanity of making dogs smoke, he outlined how Ilse Lasnitzki had already used human tissue cultures to demonstrate the harmful effects of cigarette smoke. Tissue culture was also central to a leaflet that FRAME distributed to scientists, journals, newspapers and MPs, which outlined 'the enormous potential there is for substituting animals in medical research'.[104] The possible uses that were listed for tissue culture in this leaflet, titled *Is the Experimental Animal Obsolete?*, dwarfed those of other alternatives such as computer models. Foister and Hegarty ended it by calling for the government to establish a network of tissue banks that would distribute human material to researchers.

Other anti-vivisection groups also began to promote tissue culture at the turn of the 1970s. Writing the foreword to Vyvyan's *In Pity and in Anger*, Lord Hugh Dowding, the secretary to the National Anti Vivisectionist Society (hereafter NAVS), argued that tissue culture could instantly replace three-quarters of the experimental animals used annually in Britain.[105] In 1970, the BUAV formed the Dr. Hadwen Trust to campaign for the uptake of alternative methods.[106] In 1973, the NAVS did likewise: establishing the Lord Dowding Fund to promote and fund work with alternative methods.[107] Like FRAME, these groups argued that tissue cultures were 'not being utilised to anywhere near the extent their undoubted validity warrants'.[108] The Dr. Hadwen Trust, for example, promoted tissue cultures as 'faster, more accurate, more reliable and cheaper than the inhumane and old-fashioned methods they are replacing'.[109]

Like Russell and Burch, these campaigners also argued that human tissue cultures avoided extrapolating across species barriers. To Richard Ryder, the 'overwhelming technical advantage' of tissue culture was that it 'helps avoid the dangers of extrapolating from animals to men'.[110] Some presented Thalidomide as evidence of the dangers inherent in this extrapolation, claiming that no animal test had predicted the devastating birth defects that occurred in humans. Hugh Dowding, for one, argued that 'the results of vivisection are by no means always to the benefit of humanity', and claimed that Thalidomide's side-effects would have been detected in human tissue cultures.[111]

These arguments were certainly influential. Growing numbers of politicians and newspapers began to promote tissue culture and dwelt on the limitations of animal experiments. In 1971, the Conservative MP Richard Body, chairman of a cross-party 'Humane Research Group', warned that if scientists did not promptly adopt tissue culture 'we could be heading for another tragedy on the Thalidomide scale'.[112] This was reiterated by a *Guardian* report that claimed species differences often invalidated animal experiments and called for the adoption of human tissue cultures.[113] In 1973, the Labour MP Elystan Morgan claimed that tissue culture should replace animals in all but a 'residuum' of procedures.[114] Most significantly, in 1977 the Labour Prime Minister James Callaghan declared that it was government policy to 'move to alternatives to animal experiment as quickly as possible'.[115] Callaghan claimed the use of tissue cultures to develop vaccines was 'an important advance', and stated that he would 'bend all the efforts of Departments to seeing [sic] how much further we can go in this matter'. Newspapers lent their support to the Prime Minister. The *Guardian* reasserted that tissue culture gave biologists 'unlimited amounts of standardized biological material, which is easily transported, may be kept in a freezer, is often simpler and more accurate than animal tests – and never bites'.[116]

Anti-vivisectionist groups, politicians and newspapers rarely, if ever, mentioned any of tissue culture's shortcomings. Indeed, Bernard Conyers, secretary to the Lord Dowding Fund, argued that only 'complacency' prevented scientists from replacing experimental animals with *in vitro* methods.[117] This is striking when we consider that Honor Fell had outlined tissue culture's limitations as early as 1935. It is all the more striking when we see that she made several attempts to do so again throughout the 1970s.

Honor Fell had retired as director of the Strangeways Laboratory in 1970 at the age of 70, but continued to work on the biochemistry of arthritis at the University of Cambridge's department of pathology. Nine years later, she returned to the Strangeways laboratory to continue her research on rheumatoid arthritis, and remained there until her death in 1986.[118] During the early 1970s, supporters of alternative methods often sought her advice or endorsement. In 1970, Charles Foister wrote to Fell requesting a complete bibliography of her work, to demonstrate that a scientist could succeed without using animals.[119] Fell instead sent him a long letter detailing why tissue cultures 'are not and never will be a substitute for animal experiments'.[120] As before, she claimed that tissue culture was an overly simplified environment that lacked many of the 'extrinsic factors' found in the body. Needless to say, this line of argument did not appear in any of FRAME's popular campaigning.

In 1972, Bernard Conyers also wrote to Fell for information on 'one or two aspects' of tissue culture.[121] She replied by sending a copy of a lecture she had recently given in London. Here, Fell detailed how the replacement of animals with *in vitro* methods was simply 'not practicable'.[122] She argued that while tissue culture provided information on the behaviour of cells and the localized effect of hormones, drugs and viruses, it could tell scientists:

> nothing about the physiology of an animal's circulatory or excretory systems; nothing about the physiology of its brain or its sense organs; nothing about the complex interactions of the lungs or alimentary canal. This means that a chemical compound that might appear quite harmless when tested on a tissue culture, when administered to a

patient might produce disastrous side effects that could not be reproduced in the simple *in vitro* system.[123]

Unlike Dowding and Ryder, Fell evoked contemporary concerns over drug safety to argue against the replacement of animals with tissue culture. Yet again, however, her attempts to engage with anti-vivisectionists fell flat. Despite receiving a copy of Fell's lecture, Bernard Conyers continued to portray tissue cultures as 'more humane and progressive systems of research'.[124]

In a 1976 book on tissue culture, Fell complained that: 'At the present time some well-meaning bodies are advocating the use of organ cultures and even cell cultures in place of animals. I have tried to explain the insuperable objections to this policy, but so far as I can judge without making any impression'.[125] By the late 1970s, politicians maintained that 'alternative techniques produce results which are scientifically more valid' than vivisection, and anti-vivisection groups pressed the government to adopt legislation that banned animal experiments where alternatives existed.[126] During 1978, the Labour peer Lord Douglas Houghton reminded James Callaghan of his support for alternatives, and pledged to make animal experimentation a decisive issue in the general election many believed the government would call later that year.[127]

In his 1978 book *Alternatives to Animal Experiments*, written at the suggestion of the Research Defence Society, the physiologist David Smyth sought to counter the escalating criticism of animal experimentation by outlining the limitations of supposed 'alternatives'. Smyth identified tissue cultures as the alternative 'for which the greatest claims have been made'. But he then argued that they offered a poorer model of the human body than whole animals, since they were devoid of any circulatory, digestive or nervous systems. Tissue cultures, Smyth claimed, could only provide evidence on localized phenomena and had 'very little place in studying problems in higher animals where there is interaction between tissues and organs'.[128] Contrary to the anti-vivisectionist claim that species boundaries were the major obstacle in animal experiments, he stated that 'the difference in sensitivity between cell cultures and whole animals of the same species may be far greater than the difference between the whole animal of two different species'.[129]

Smyth also claimed that it was misleading to frame Thalidomide as evidence of the dangers in extrapolating across species, since the issue here 'was nothing to do with species variation'.[130] He detailed how the drug's harmful effect on developing foetuses had gone undetected because traditional methods of toxicity testing did not use pregnant animals. Testing Thalidomide on male and non-pregnant humans, or on human tissue cultures, would still not have uncovered the drug's side effects, but tests on pregnant mice or guinea pigs would have done. All this proved, Smyth argued, was that all available tests were necessary to determine a drug's side effects. He concluded that since they said nothing about the systemic effects of compounds, testing a drug solely on tissue cultures would be tantamount to 'criminal irresponsibility'.[131]

Smyth drew upon the opinions of many scientists to show that his book was not 'subjected to approval or censorship by the Research Defence Society'.[132] In particular, he cited long passages from the introduction Honor Fell wrote to a 1976 book on tissue culture, where she refuted the claims of anti-vivisectionists. Smyth claimed that Fell's criticism carried the most significance, since she had 'a personal interest in the value of tissue culture'.[133] Replicated in Smyth's book, which was endorsed by the MRC and covered by several newspapers, Fell's arguments finally began to carry some weight. The *Guardian*, for example, outlined how 'the possibilities for alternatives are more restrictive than some people make out'. The ultimate test of a new compound, it explained, 'has to be on the whole animal, not on a few cells in a dish, because only the whole animal can identify the weak link in the body's chain'.[134]

Despite the efforts of groups such as FRAME, animal experimentation was not a major issue in the 1979 general election. Economic problems and trade union strikes during the so-called 'winter of discontent' dominated the political landscape; and Margaret Thatcher's Conservative Party won the election on a pledge to curb union power and correct the economy, not to promote substitutes for experimental animals. Although groups such as FRAME, the Lord Dowding Fund and the Dr. Hadwen Trust continued to publicize tissue culture during the 1980s, they no longer presented it as a failsafe alternative to animal experiments. These claims, which were influential during the 1970s, were gradually undermined by the scientific discussion of the differences between cultured cells and whole organisms.

Conclusions

Histories of biology during the 1960s and 1970s generally focus on the development of organ transplants, genetic engineering or *in vitro* fertilization, and overlook how the discussion of species barriers was a prominent part of the scientific and popular landscape in this period.[135] As we have seen, the creation of human–animal hybrid cells in 1965 was a significant part of the 'biological revolution', both for fostering a sense of radical power and for encouraging ambivalence towards science. Between 1965 and 1967, these hybrid cells appeared in public as regularly as 'test-tube babies' – in newspapers, popular books and on television. As before, this public profile was largely due to the material practices and bold claims of the scientists who created them. Harris and Watkins deliberately played on fascination with species differences and the 'biological revolution' to gain attention for cell fusion, deliberately fusing antithetical species and publicly predicting hybrid organisms. Similarly, few historians have examined how the anti-vivisection movement gained legitimacy in the 1970s by promoting human tissue cultures, and specifically claiming that they effaced the species barriers that made animals poor models of the human body. Some campaigners for alternative methods, like Charles Foister and Richard Ryder, were scientists themselves, and their presentation of tissue culture as superior to vivisection clearly influenced many MPs and journalists.

But other prominent scientists, like Honor Fell, rejected this presentation and argued that cultured cells were poorer models of the human body than animals of a different species. This formed part of a long-standing critique that questioned tissue culture's accuracy as a 'proxy body'.[136] It dated back to 1910, when Justin Jolly claimed that Alexis Carrel was merely witnessing cell death *in vitro*, and it was also evident in the interwar belief that cultured cells regressed to a 'primitive' state that was unrepresentative of the body. But the form this argument took was itself culturally mediated. The interwar emphasis on de-differentiation reflected contemporary concerns over degeneration, and played out in popular portrayals of atavistic tissue cultures. By the 1960s and 1970s, in order to refute images of monstrous chimaeras and a growing anti-vivisection movement, scientists now emphasized the fundamental differences between cells and animals, and stressed the importance of the whole organism.

6
Nobody's Thing? Consent, Ownership, and the Politics of Tissue Culture

As we have seen, the scientific collection of tissue for research was widely viewed as unproblematic for much of the twentieth century. But this practice became contentious during the 1970s and 1980s, when excised tissues became the subject of often heated debate, with scientists, social groups, lawyers and a new breed known as 'bioethicists' questioning the ethics and legality of the procedures that transformed them into experimental tools. These questions played out in academic conferences and journals, in court, in bioethical reports and government legislation, in newspapers and even spilled onto the streets in protests.

The new scrutiny of research on tissues reflected historically specific concerns. For instance, unease to research on foetal tissue culture during the 1970s reflected pro-life opposition to the recent liberalization of abortion laws. The discussion of research on foetal tissues prompted debates regarding whether scientists should obtain informed consent for tissue samples, and whether patients might be entitled to own tissues, which reflected the political and bioethical belief that patients were autonomous consumers of healthcare. Allowing an individual to determine the fate of excised tissue samples seemed to many a laudable extension of this new ideology.[1] Indeed, at a more general (though no less relevant) level, this stress on patient autonomy reflected new neo-liberal governmental practices where 'active' citizens were encouraged to engage with and reshape public sector services, including medicine.[2]

Charting these debates further undermines any dichotomy between scientific and social concerns. Some of the earliest advocates of patient consent and ownership were scientists, while some lawyers and ethicists opposed these new measures on the grounds that they contravened tissue's common-law status as *res nullius* – 'nobody's thing'. Furthermore, surveys of public opinion revealed a marked ambiguity that mirrored differences of opinion within science, law and ethics. There was no consensus on this issue in any professional or social group and, as throughout this book, we see considerable interaction between them.

Pro-life politics and foetal tissue culture

The collection of human tissues for research had not featured in the growing criticism of biology during the late 1960s. Indeed, human tissue culture was often represented positively in this period, thanks to the anti-vivisectionist promotion of alternatives to animal experiments. Positive attitudes to work on human tissues were evident in 1970, following Norman St. John-Stevas's claim that doctors kept aborted foetuses alive for research purposes and encouraged abortions in order to provide scientists with foetal tissues.[3] Stevas was a Catholic and Conservative MP, and was attempting to instill opposition to the 1967 Abortion Act, which had legalized abortion during the first 28 weeks of pregnancy when two doctors vouched for its medical or psychological necessity.[4] But newspapers did not share his unease, and instead focussed on the positive aspects of research on foetuses and foetal tissues. While *The Times* acknowledged that readers might be troubled by the research on aborted foetuses, it stressed that such work was essential to the development of incubators for premature babies.[5] Others went further when outlining the benefits of foetal tissue cultures. The *Daily Express* detailed how they had been central to the development of polio and rubella vaccines, which saved thousands of lives, and argued that research on foetal tissue cultures was 'essential *and causes no concern*'.[6]

Following Stevas's claims, the government convened an inquiry into research on foetuses and foetal tissues, chaired by the gynaecologist Sir John Peel. The committee's report, published in 1972, also saw no problem with research on foetal tissues, or how scientists obtained them. Detailing how vaccines for polio and rubella were produced on foetal tissues, it claimed that 'the use of foetal tissues has gone beyond basic research into the field of preventative medicine'.[7] The report then outlined how foetal tissue was the only adequate material for research on pathogens like the cold-causing rhinovirus, which 'do not grow on cultures of non-human cells'.[8] Like the press, Peel's report argued 'there was no reason to object to the use of foetal tissues' in research (but it stressed that 'in no circumstances should there be monetary exchange for foetuses, foetal tissue or foetal material').[9] It also stated that where the foetus was removed in operations that led to 'the termination of its life' – i.e., abortions – there was 'no statutory requirement to obtain consent for research'.[10] The report argued that seeking additional consent for research on foetal tissues would 'be an unnecessary source of distress to parents', although it did urge doctors to respect the wishes of parents who expressed 'special directions about the disposal of the foetus'.[11] Here, the Peel committee treated aborted foetuses as identical to other materials removed during a clinical procedure. Like gall bladders, tumours and biopsies, they attained the status of surgical waste on removal from the body and could legitimately be transferred to researchers.[12] As was the case throughout the twentieth century, consent to a clinical procedure was taken to imply abandonment of tissue and, unless they stated otherwise, the patient gave up any right to determine what happened to it.

This categorization reaffirmed the common law view of excised body parts as *res nullius* – 'nobody's thing'. A line of cases dating back to the seventeenth century ruled that human body parts could not be owned: meaning no-one had standing to bring a claim for damage or theft of human material, and that tissues could not be dealt with under the usual rules of inheritance or exchange.[13] However, this position was subject to an important exception, deriving from the 1908 Australian case *Doodeward v Spence*, which centred on the ownership of a two-headed foetus that had been displayed as a curio and then confiscated on the grounds of public indecency. Ruling in favour of the individual who preserved the foetus in formalin and displayed it, the judge held that:

> When a person has by lawful exercise of work or skill so dealt with a body or part of a body ... that it has acquired some attributes differentiating it from a mere corpse awaiting burial, he acquires a right to retain possession of it.[14]

This 'work-and-skill' principle drew on John Locke's approach to the notion of property, which granted an individual ownership of natural material if 'he hath mixed his Labour with [it] and joined it to something that is his own'.[15] While no-one questioned the legal status of tissue cultures for much of the twentieth century, it would appear that scientists felt bound by this convention. As a researcher who collected tissue between the 1940s and 1970s attested, if you successfully cultivated a sample 'then it was yours'.[16]

A similar state of affairs had long existed in the United States, where excised tissues were also regarded as surgical waste and distributed to researchers. But during the 1970s, this arrangement was subjected to unprecedented scrutiny and criticism. In direct contrast to Britain, this criticism first arose in debates concerning foetal tissue cultures. This difference was primarily due to two factors. For one, abortion was a far more politicized issue in America than in Britain. While British pro-life groups were politically marginal during the 1970s, their American counterparts were well funded, politically influential and launched a sustained attack on the liberal abortion laws that were passed following the 1973 case *Roe v Wade*, where the Supreme Court ruled that a woman's constitutional right to privacy included the right to have an abortion in the early stages of pregnancy (and later, if continuing the pregnancy endangered her health).[17]

Another crucial difference was the increasing scrutiny of American biomedicine from academics in other fields, known as 'bioethicists'. While Maurice Pappworth's 1967 book *Human Guinea Pigs* had drawn attention to the ethics of human experimentation in Britain, the governance of clinical or basic research had not been redefined as a matter that required external input. Evoking the emphasis on clinical autonomy struck as part of the 1948 settlement that created the National Health Service, doctors successfully argued that ethical responsibility should remain 'firmly on the shoulders of the medical profession'.[18] This was evident in Peel's report, which asserted that 'ethical questions are for the profession to consider'.[19] In the United States, however, following an outcry over experiments on institutionalized children in New York, and the intentional withholding of syphilis treatment from black Americans in Tuskegee, Alabama, growing numbers of theologians, lawyers, sociologists and philosophers were appointed to federal investigations into biomedical research, and became public authorities on scientific and medical ethics.[20] In the wake of *Roe v Wade*, some bioethicists, like the Methodist theologian Paul Ramsey, claimed that foetal research posed ethical problems that had been too quickly dismissed in Britain. Ramsey argued that evoking medical benefits was insufficient justification for research on aborted foetuses, and recommended they be afforded the same consideration as vulnerable or dying patients. Medical progress, he cautioned, was not 'the sole source of medical ethics'.[21]

Tellingly, Ramsey also detailed how anti-abortion campaigners in the United States regularly insisted that 'research on fetal tissues is as outrageous as research on the whole living fetal human being'.[22] By the mid 1970s, pro-life groups and some congressmen argued that research on foetal tissues should be banned, irrespective of its medical benefits, since it entrenched acceptance of abortion among the medical community and society at large.[23] Activists even picketed laboratories that were known to work on foetal tissue cultures, such as Philadelphia's Wistar Institute. Since the late 1950s, Wistar researchers had used foetal tissue cultures to propagate different viral strains – circumventing strict abortion laws by importing material from the Karolinska Hospital in Sweden, where abortion was permitted on eugenic, psychiatric, medical and socio-economic grounds.[24] By the mid 1960s, the young biologist in charge of this work, Leonard Hayflick, had established over 25 different foetal cultures, labelling the most stable and well characterized as 'WI-38'. Along with Hillary Koprowski, the head of the Wistar Institute, Hayflick publicly endorsed WI-38 as the best substrate for vaccine production, claiming it was free from the viruses that lurked in monkey and adult tissues.[25]

Although they encountered resistance in the United States, where the Nobel-winning virologist Albert Sabin stressed the merits of monkey tissue, Hayflick and Koprowski successfully lobbied policymakers at the MRC, who ordered over 250 ampoules of WI-38 for the production of polio vaccine in Britain in 1965.[26] Keen to increase the available stocks of foetal tissue, and to avoid reliance on imports of WI-38, the MRC commissioned the development of a comparable foetal culture the following year. This culture, labelled MRC-5, was derived from the lung tissue of a foetus aborted on medical grounds at St. Bartholomew's Hospital and was promptly used to develop hepatitis and rubella vaccines.[27] The growing enthusiasm for these foetal cultures was by no means confined to Britain. As the 1960s progressed, scientists and pharmaceutical firms across Asia, Africa and Europe also began to demand WI-38; and like George Gey before him, Hayflick became known for personally delivering stocks of WI-38.[28] Hayflick and Korprowski's dogged promotion of WI-38 in the United States finally paid off in 1973, when the Charles Pfizer Laboratories received a licence to develop and sell a polio vaccine developed on WI-38.[29]

At the same time, Hayflick incorporated WI-38 into research on the biology of aging after noticing that its cells invariably died after around 50 divisions. This observation challenged Carrel's long accepted claim that cultured cells were immortal, with any death *in vitro* due to either infection or bad technique. It appeared so surprising that the *Journal of Experimental Medicine*, which had published Carrel's paper on the 'old strain', rejected Hayflick's work on the grounds that 'the largest fact to have come out of tissue culture is that cells inherently capable of multiplying will do so indefinitely if supplied with the right milieu *in vitro*'.[30] But support for the so-called 'Hayflick limit' grew once other scientists demonstrated that non-cancerous cells did die after around

50 cell divisions; and biologists eventually accepted that they had a finite lifespan *in vitro*.[31]

With WI-38 increasingly central to vaccine production and research into aging, or 'gerontology', the US National Institute of Health (hereafter NIH) provided the Wistar Institute with a grant solely for its production, storage and distribution during the early 1970s. But this coincided with pro-life opposition to research on foetal tissues. As well as picketing the Wistar Institute, pro-life activists protested at the high-profile launch of the Skylab III satellite in November 1973, after learning from newspapers that WI-38 was being fired into space to investigate the cellular effects of space travel.[32] These protests escalated the following year, when campaigners in Massachusetts successfully evoked an 1814 grave-robbing law to secure an indictment against three scientists who obtained an aborted foetus for research. Despite protestations that no ethical guideline had been breached, the assistant District Attorney maintained that 'had the researchers asked each woman ... for permission to perform what amounts to an autopsy on her dead, aborted foetus, there would be no case'.[33] Although the doctors were acquitted, scientists began to question whether it was now a criminal offence to obtain foetal tissue without seeking parental consent.[34] Their uncertainty was compounded in July 1974, when President Richard Nixon signed the National Research Act that resulted from congressional hearings on the ethics of human experimentation. Following pro-life pressure, the Act temporarily banned research on any living foetus obtained by abortion.[35] But this legislation did not define what it meant by 'living'. As a result, many scientists stopped using foetal tissue cultures – fearing reprisals for using tissue that came from abortions, which could be said to exhibit some form of 'life' *in vitro*.[36]

When the ban on foetal research was lifted in August 1975, NIH guidelines recommended that researchers obtain parental consent before undertaking research on the aborted foetus.[37] This was partly an attempt to prevent another indictment for non-consented 'grave-robbing'; but it also reflected the increasing emphasis on individual autonomy, and demands for full disclosure of procedures and risks, that featured in debates on the rights of patients and experimental subjects. In 1974, President Nixon responded to controversies over human experimentation by establishing a fixed-term National Commission for the Protection of Human Subjects in Biomedical and Behavioural Research. The Act that established the Commission stipulated that no more than five of its 11 members should be doctors or scientists – with the majority drawn from philosophy, law, theology, sociology and the general public. The Commission's recommendations, issued as the *Belmont Report*, made respecting patient autonomy a guiding principle for all biomedical researchers.[38] Amidst this heightened emphasis on patient autonomy, the previously uncontested fate of excised tissues, and their status as *res nullius*, was increasingly challenged. In a 1974 case, *Browning v Norton Children's Hospital*, a patient with a pathological fear of fire brought an action against the surgeon and hospital for the mental anguish that he allegedly suffered on learning that his amputated leg had been incinerated as surgical waste.[39] The following year, in *Mokry v University of Texas Health Science Center in Dallas*, the plaintiff claimed to have suffered nervousness and distress after learning that his eyeball was accidentally lost down a plughole during a pathological examination.[40] As the lawyer Bernard Dickens argued in 1977, these cases demonstrated that patients increasingly appeared to be taking an 'interest in the use and disposition of their tissues'.[41]

American scientists often turned to lawyers and philosophers to prevent their work coming under fire during the 1970s, and tissue culture was no exception.[42] Several philosophers and lawyers attended a 1976 meeting of the Tissue Culture Association, where the academic lawyer William J. Winslade urged scientists to seek consent before working on any excised tissues – either by approaching patients directly or distributing forms to the attending physician. Winslade stated that questions were certain to be asked of tissue culture 'now that the legal establishment has begun to scrutinize the scientific community's work on human subjects'.[43] He recommended that scientists should adopt informed consent as a 'minimum requirement', in order to make tissue culture research 'legally permissible and morally sensitive'.[44] Echoing the broader stress on patient autonomy, Winslade continued that 'requiring informed consent for tissue not only removes the taint of impropriety stemming from non-disclosure, but also gives a person an opportunity to express his/her desires and provides for the values of privacy and self-determination'.[45] Winslade framed informed consent as beneficial to science, arguing that once patients were informed about the fate of excised tissues and the 'legitimate purpose of research', it was likely that 'most ... would freely consent to contribute their tissue'.[46] However, he warned that if tissue culture researchers did nothing but 'wait until criticism, litigation or bad publicity forces them to take action, it is likely that research will be greatly impaired'.[47]

Some American scientists dismissed such proposals as 'foolish restrictions on research'.[48] Others agreed with giving patients the right to control what was done to their bodies in treatment and research, but dissented when this extended to tissue samples. One Tissue Culture Association member argued that while tissues may have originated as part of a human being, their removal from the body meant they were no longer 'properly speaking, parts of persons'.[49] Another claimed that obtaining fully informed consent for research on tissues was impossible, since neither the patient nor the scientist 'can appreciate the possible consequences or ways in which tissues may be used sometime in the future'.[50] Some, too, envisaged a situation where patients or research subjects went so far as to claim full ownership over excised tissue samples, demanding the right to control their future use in research as well as, perhaps, being able to sell them to scientists. One Tissue Culture Association member stated it had become 'quite unclear who has rights in cut hair, nail clippings and cells taken from human bodies', and added that 'we are in a state of legal and moral confusion when questions are raised about whose permission is required for the use of material obtained in these ways'.[51]

Winslade argued that granting patients ownership of tissue samples was not inconceivable, as blood had been considered personal property in the United States since the 1954 decision in *Perlmutter v Beth David Hospital*, and was part of a thriving market that also included 'urine, hair, fingernails, etc'.[52] Yet he also acknowledged that this commercial model raised more problems than it promised to solve. All the bodily materials currently bought and sold were regenerative; and different rules would be needed to limit the sale of non-regenerative organs and tissues, and to prevent the exploitation of financially disadvantaged donors by third-party agents. In the meantime, Winslade asserted, new consent protocols remained essential.

Who owns WI-38? Tissue culture goes to court

One delegate at the Tissue Culture Association meeting warned fellow attendees that while ownership was not yet a pressing issue, 'some of my colleagues and myself have the feeling that [it] will become a significant problem over the longer term'.[53] Events later in 1976 vindicated this prediction, with the onset of the first lawsuit involving ownership of a tissue culture. Given that questions surrounding consent and ownership arose largely due to pro-life unrest, it is fitting that this unusual custody battle centred on WI-38.

In February 1976, Leonard Hayflick resigned as professor of microbiology at Stanford University – prompted by a growing struggle over the large stocks of WI-38 he had brought with him from the Wistar Institute. When Hayflick was appointed to Stanford in 1968, the NIH had asked him to take only ten ampoules of WI-38 and leave the remaining 400 ampoules at the Wistar Institute. However, as an article in *Science* noted, Hayflick 'took the lot to Stanford anyway'.[54] He then set up a private company, Cell Associates Inc., to sell WI-38 to researchers and pharmaceutical firms. In 1975, after applying for the position as head of the new National Institute of Aging, Hayflick asked the NIH to clarify the ownership status of WI-38. After auditing Cell Associates, the NIH produced a damning report that alleged Hayflick had made over $67,000 by selling federal property. The NIH confiscated all stocks of WI-38 and, under threat of dismissal, Hayflick resigned from Stanford. The story made the front page of the *New York Times*, in a report that portrayed Hayflick as a racketeer who had stolen and sold 'property of the federal government'.[55]

The NIH claimed that it was the rightful owner of WI-38 because it had funded the research in which the culture was established, and then provided the Wistar Institute with a grant solely for its maintenance and distribution. Hayflick, on the other hand, argued that he was the rightful owner since he had performed the crucial task of transforming the foetal tissue into WI-38. Stating his case in *Science*, he claimed that 'I felt, and am justified in feeling, that these cells are like my children'.[56] In the first move of what was to become a protracted legal battle, the NIH launched criminal proceedings against Hayflick for stealing federal property. Hayflick countered by asserting his rights to WI-38, and suing the NIH for defamation. During five years of hearings, Hayflick and his legal team notably argued that WI-38 might also be considered the property of the Swedish couple whose aborted foetus originally provided the tissue. This was the first time that the possibility of patient or family ownership had been raised in court. As Hayflick later wrote: 'We argued, I believe for the first time, that not only did my former institution and I have a legitimate claim to these cells, but a good case could also be made for title to be vested in the parents or estate of the embryo from which WI-38 was derived'.[57]

The case was eventually settled out of court in September 1981. The NIH retained possession of most of the WI-38 stocks it confiscated in 1976, while Hayflick retained his profits from Cell Associates and the right to continue selling WI-38.[58] The settlement arose largely because Hayflick's case had been strengthened by legal changes that occurred while the lawsuit was in progress. In an attempt to encourage private investment in biotechnology, the US Court of Customs and Patent Appeals had argued in 1977 that there was no justification for excluding the creator of genetically modified microorganisms from the patent system. Then, as part of efforts to 'combat a slump in US productivity and effectiveness' in 1980, the newly elected President Ronald Reagan signed the Bayh-Dole Act, which granted individual scientists and universities the right to patent inventions made in the course of federally funded work.[59] The same year, in the case *Diamond v Chakrabarty*, the United States Supreme Court was called upon to determine whether the legal definition of patentable inventions could be interpreted to include genetically modified organisms. This case saw the courts dealing with a scientist who argued that genetically modified bacteria were patentable inventions due to being discontinuous from any natural state. The court agreed, holding that patent legislation should be interpreted to include organisms and biological materials that had been altered through human ingenuity.[60] This definition crucially meant that a scientist who cultivated tissue could be considered the rightful owner of, and would be allowed to profit from, tissue cultures such as WI-38: transforming what the NIH originally viewed as theft into what Hayflick celebrated as 'praiseworthy federal policy'.[61]

But while these rulings may have settled whether or not altered biological materials were patentable, and whether or not scientists could profit from them, they did not prevent excised tissues becoming embroiled in ethical and legal disputes. The shifting economic climate that proved fortunate for Hayflick was to raise more problems than it solved. As the journalist Barbara Culliton noted in 1984, while questions surrounding patient consent and ownership had been posed in the 1970s, their time had truly come 'in this new era of commercialization'.[62]

Commercialism and its discontents: Struggles over consent and ownership in the 1980s

Because of these regulatory changes, human tissues quickly attained considerable financial potential. A 1987 report by the United States Office of Technology Assessment detailed how the use of human tissue in medical research had more than trebled in the four years after the *Diamond* case. It also outlined how patents on products derived from human tissues amounted to 20 per cent of the total patents filed by medical schools between 1980 and 1984.[63] During the early 1980s, the European Patent Office also decided that modified biological materials were patentable inventions, and issued 235 patents for human tissue cultures and their derivative products between 1982 and 1994.[64]

But these changes raised significant problems. Less than five months after the *Diamond* case, another lawsuit came to light which demonstrated, to *Science*, how 'the powerful forces of the profit motive clearly have the capacity to strain and rupture the informal traditions of scientific exchange'.[65] This case involved a cell line known as 'KG-1', which Phillip Koeffler and David Golde, haematologists at the University of California, Los Angeles (hereafter UCLA), had created by cultivating lymphoma cells from a terminally ill patient in 1977.[66] As was common practice, Golde sent a sample of the cell line to Robert Gallo, a researcher at the National Cancer Institute who wanted to screen for any possible cancer-causing viruses. Gallo noticed that KG-1 produced interferon, a protein known to inhibit viral replication, and passed another sample to Sidney Pestka, who worked at the Roche Institute of Molecular Biology, which was a research arm of the pharmaceutical giant Hoffman La Roche.[67] By manipulating the cell cycle and modifying the culture medium, Pestka prompted KG-1 to produce significant amounts of interferon. Hoffman La Roche then contracted researchers at Genentech, a new biotechnology firm, to isolate and purify the interferon gene from KG-1. Genentech then announced plans to publicly list one million shares at around $30 each. As *Science* noted, the timing of this announcement was no coincidence, as the value of Genentech's shares rested 'heavily on its chief potential money spinner – interferon made to the instructions of the KG-1 gene'.[68]

In order to safeguard potential revenues, Roche and Genentech filed a joint patent to cover both the interferon produced by KG-1 and the method used to isolate genes from the cell line. Bertram Rowland, a patent attorney who represented UCLA, immediately countered that Roche and Genentech had 'subverted' the traditional free exchange of tissue samples, and claimed Golde had not given Pestka permission to distribute KG-1 to anyone else.[69] Rowland argued that UCLA, as home of the scientists who originally created KG-1, should be seen as the rightful owner that was entitled to profits from the sale or licensing of inter- feron derived from the cell line. On 12 September 1980, Roche sought resolution by filing suit in a California court. The lawsuit was eventually settled out of court in 1983, with Roche and Genentech retaining the right to market interferon produced by KG-1, in exchange for payment of undisclosed royalties to UCLA.[70] But as an article in *Science* queried, 'mightn't the [interferon] gene be regarded as the unalienable property of the individual to whom it belonged, or, since he is dead, to his heirs?'[71] Neither side, *Science* noted, seemed willing to countenance this possibility as one of the 'ground rules that have to be worked out'.[72]

The issue of patient or familial ownership featured prominently in the next dispute over a tissue culture, which broke just as the KG-1 case was resolved. This centred on a hybridoma cell line, which is generated by fusing an antibody-producing β-cell to an immortalized cell *in vitro*.[73] The hybridoma in question was created when Hideaki Hagiwara, a visiting postdoc at the University of California, San Diego (hereafter UCSD), provided colleagues with β-cells taken from his mother, who was suffering from cervical cancer. After fusing these cells with an established cell line, Ivor Royston, a UCSD oncologist, discovered the resulting hybridoma secreted high levels of antibodies that impaired the growth of cancer cells from the lung, colon, prostate and cervix.[74] To Royston's surprise, Hagiwara then took the hybridoma cells to Japan, arguing they were family property since one-half originated from his mother. While Hagiwara claimed he wanted the hybridoma to treat his mother's cervical cancer, Royston and other UCSD scientists pointed out that Hagiwara's father was president of the Hagiwara Institute of Health, a private company, and questioned whether he had 'spirited the cells away to exploit them commercially'.[75] They argued that property rights rested with UCSD, on account of the 'art and expertise' involved in creating the hybridoma.[76]

Both sides quickly reached agreement once this latest ownership tussle became public. The Hagiwaras were embarrassed by growing accusations of theft, and UCSD staff were keen to avoid a repeat of the protracted KG-1 case. The resolution assigned patent rights to UCSD and gave the Hagiwara Institute of Health exclusive license to market the hybridoma in Asia, on the condition they paid UCSD if it proved commercially valuable.[77] Although Hagiwara's claim was not successful, it prompted greater discussion of what property rights patients and their families had in cultured tissues. Crucially, there was no consensus among scientists or lawyers on the issue. Bertram Rowland, who represented UCLA in the KG-1 case, believed the question was 'irrelevant' because a hybridoma was 'a newly created biological entity' and therefore belonged to the scientists who created it.[78] While Ivor Royston agreed, he argued that consent forms should be modified to give patients the opportunity to determine whether excised tissues could be used, stating they would not be entitled to compensation if research proved lucrative. Royston believed this would pre-empt future claims to patient ownership, since patients who objected to research could withhold consent, while those who consented waived any future claim to ownership.[79] By contrast, Gordon Sato, another UCSD scientist, hinted at recognition of patient ownership when he recommended that patients should automatically be given a share of any profits that may follow from research on their tissues. Harold Jackson, Hagiwara's attorney, similarly claimed that tissue cultures should not automatically be considered scientific property on account of discontinuity from a 'natural' state. The ownership of tissue cultures or cell lines, he followed, was 'an area of law that needs to be explored'.[80]

Nevertheless, the issue of patient ownership had not yet been considered by the American courts. This changed in 1984, when John Moore, a Seattle businessman, filed a suit against scientists at UCLA (including David Golde) who removed his spleen during treatment for leukemia in 1976, then converted it into a commercially valuable cell line without his knowledge or consent. Golde and colleagues became aware that this cell line, known as 'Mo', produced high quantities of interferon and signed lucrative contracts with biotechnology companies. Moore became suspicious in 1983 when Golde, perhaps as a consequence of the Hagiwara case, sent him a consent form that would grant UCLA rights to all products derived from tissue obtained during his operation and follow-up visits. Moore refused to sign the form and contacted an attorney, who alerted him to the Mo cell line – then estimated to be worth $3 billion.[81] The central claim in the subsequent case, *John Moore v the Regents of the University of California*, was the unauthorized conversion of Moore's spleen tissue into the Mo cell line. 'Conversion' is a tort or civil wrong that occurs when a person deals with property that does not belong to them, in a way that is inconsistent with the rights of the lawful owner. By charging UCLA scientists with conversion, Moore claimed a possessory right to his excised spleen cells. But he wanted more than control over, or the return of, the Mo cell line. Fully aware of its commercial potential, and of the contracts Golde had already signed, Moore also argued that he had a right to profit from the commercial development of the cell line and its downstream products.[82]

From the outset, scientists, lawyers and ethicists framed *Moore* as the embodiment of the problems caused by the heightened commercialism in biomedical research. Reporting on the case for *Science*, Barbara Culliton asked if 'in this new era of the commercialization of biological science, in which people have grand (and exaggerated) visions of making money, should persons who donate tissue to research be given some contractual right to share in any profits that might one day ensue?'[83] In 1985, the Dean of Medicine at Yale University did likewise, equating Moore's stance with that of a general 'public' and stating that:

> The public cannot help but see that the goals of some scientists
> – clinical or basic – are different than in the past. No longer can the
> biological scientist simply be portrayed as a dedicated, noble, under-

paid truth-seeker trying to unlock nature's secrets for humankind's benefit. The biotechnology revolution has moved us, literally or figuratively, from the class room to the board room and from the *New England Journal* to the *Wall Street Journal*. ... Small wonder then that a patient who sees his or her tissues becoming commercially valuable might say 'why shouldn't I share in the profits? After all Dr. X wouldn't be getting rich without me'.[84]

Most coverage pointed to *Moore*'s possible consequences. In the *Wall Street Journal*, the financial journalist Alan Otten warned that granting patients ownership over tissue would derail much 'potentially valuable research'.[85] Otten predicted that scientists would spend valuable time and money buying material from patients, and maybe even sharing profits. But he noted that research could just as likely be impeded by a growing public belief that patients were being unwittingly exploited by researchers and pharmaceutical firms. If this sense grew there was, he continued, a real danger that the public willingness to co-operate in research would evaporate.

Some scientists framed *Moore* as 'outrageous' and a 'threat to the sharing of tissue for research purposes'.[86] This was evident in a *Nature* article, which carried an illustration depicting an avaricious lawyer interrupting an operation to demand the patient be granted full rights to his tumour (see Figure 6.1).[87] Others, however, endorsed some form of patient ownership in tissue. As Culliton reported, 'several researchers ... see no reason why some sort of provision should not be made that would give patients rights in cases such as this'.[88] Ivor Royston, for one, demanded 'new laws which delineate the rights, if any, of patients to commercial products of cell lines derived from their tissues'.[89] Royston suggested that where a patient was not known to an investigator, and where research on tissue did not identify the patient or lead to commercial products, it was sufficient to obtain their informed consent for research. And in the event that the patient or family was aware that research on their tissues may lead to development of commercial products, he reiterated that consent forms should contain a sub-section asking the patient to waive their rights to the tissue and its downstream products.

This lack of consensus was mirrored by diverse legal and bioethical positions. In 1986, Lori Andrews, a pro bono member of John Moore's legal team, wrote in the bioethical *Hastings Center Report* that patients 'should have the autonomy to treat their own parts as property'.[90] Framing *Moore* as evidence of a growing demand for control over excised tissues, and for patient consumerism generally, Andrews stated that new

Figure 6.1   Nature represents the *Moore* case as an imposition onto biomedical science. Note the contrasting portrayals of the humanitarian doctors and the sinister lawyer. Reprinted by permission from Macmillan Publishers Ltd., Sandra Blakeslee, 'Patient Sues for Title to Own Cells', *Nature*, Vol. 311 (1984), p. 198.

rules were needed to acknowledge that 'people have an interest in what happens to their extracorporeal body parts'.[91] But the extent to which *Moore* reflected a broad popular opinion was questionable. To support her argument that people cared about the fate of tissue, Andrews cited an opinion poll where only 20 per cent of respondents expressed unease at its use in research.

Nevertheless, Andrews called for the adoption of a free-market model where patients were considered the owners of tissue and were able to sell it to scientists and doctors. She rejected the view that this would encourage the exploitation of the poor and vulnerable, who might be coerced into selling tissues and cells. Banning payment on these grounds, she claimed, overlooked the fact that 'to the person who needs money to feed his children or to purchase medical care for their parent, the option of not selling a body part is worse than the option of selling it'.[92] Andrews also disputed that granting patients ownership would burden scientists with negotiation and bartering. If a patient were reluctant to sell his or her tissues, she continued, the researcher could simply approach someone who was not.[93]

The bioethicist Arthur Caplan rejected this free-market model, claiming it would threaten traditional doctor-patient relations, and that the exorbitant prices individuals might command would drastically reduce the supply of tissue. He claimed that the solution to *Moore* instead lay in full scientific disclosure to patients of an interest in excised tissues. 'When research is the goal', Caplan stated, 'whether for profit or not, those whose materials are to be used in research have a right to know'.[94] The final judgement in *Moore* reflected Caplan's stance. In 1990, the California Supreme Court took the broadly Lockean view that there was no basis for conversion since the Mo cell line was 'factually and legally distinct from the cells taken' (this overturned a 1988 judgement by the California Court of Appeal, which ruled that Moore had grounds for conversion).[95] The court did, however, hold that Golde and other UCLA staff were liable for failing to disclose an interest in producing a cell line from Moore's spleen.

Writing in the *New England Journal of Medicine*, the lawyer William J. Curran claimed the Supreme Court decision had closed debates on patient ownership.[96] But this was wishful thinking. *Moore* was only binding in California and there was nothing stopping similarly aggrieved patients bringing suit elsewhere. Other ethicists and lawyers saw future claims as inevitable, and continued to debate whether patients should be granted property in excised tissues or, at the very least, should be asked to give consent before they were used in research.[97]

Britain: Debating policy and surveying public opinion in the 1980s and 1990s

This discussion was not confined to the United States. In Britain, where little attention had been paid to the collection of tissues since the 1972 Peel report, *Moore* prompted an often anxious debate regarding the ethical and legal status of excised material. By the late 1980s, the British landscape appeared to be changing in ways that encouraged a similar case to *Moore*. Medical practices came under increasing fire from public figures like the academic lawyer Ian Kennedy, who regularly praised the 'brilliant insights' of American bioethicists and called for greater patient involvement in the development of regulatory guidelines.[98] Kennedy's argument dovetailed with the neo-liberal ambitions of Margaret Thatcher's Conservative government, who believed that many professions, including medicine, should be remodelled on market lines and made accountable to newly-empowered end users.[99] Like the Reagan administration in the United States, Thatcher's government also encouraged greater commercial incentives in biology, ensuring that human tissues were increasingly used in lucrative research.[100] Taking stock of these changes in a 1988 piece for the *Lancet*, the barrister Diana Brahams noted that the collection and research on tissues taken from living patients was not covered by the 1961 Human Tissue Act, and may be ruled as 'neither ethical nor lawful' if challenged in a British court.[101] Brahams argued that tissue's status as *res nullius* was less tenable thanks to the growing emphasis on empowered patients, and to the fact that scientists could use genetic analysis to obtain personal data from tissue samples. In the light of these factors, she argued, it was 'unlikely that a patient being operated on can be said to have abandoned (and lost any rights in) his organs'. Brahams also noted that recent convictions for theft of other supposedly 'abandoned' items, such as dustbin refuse and lost golf balls, set a worrying precedent.[102] She predicted there may be nothing to prevent a British judge ruling that a patient 'might well be entitled to all the proceeds produced by the cell line less the cost of developing and maintaining it'.[103]

British groups with a vested interest in tissue culture sought to pre-empt this possibility by calling for the adoption of new practices and legislation. In the 1987 edition of a tissue culture manual, the Glasgow oncologist Ian Freshney urged researchers to obtain the written consent of patients before they collected tissue.[104] The same year, FRAME argued that continued public confidence in non-animal methods depended on the promulgation of a 'sensible and widely adopted' policy to regulate the acquisition of human tissue.[105]

In order to determine what form this policy should take, FRAME sent 1,000 copies of a questionnaire to names and addresses drawn from the electoral register. Recipients were asked whether they would consent to research on their tissue; whether they preferred its use in medical, industrial, or cosmetic research; the sorts of tissue they would be willing to donate; and their views on animal experiments. But the results were inconclusive, undermining any hope of clear policy guidance. Only 200 people responded, and only 53 per cent supported use of their own tissue. FRAME questioned whether these replies were even representative of public opinion. Written comments suggested that replies mainly came from individuals with strong views on vivisection; and these respondents were disproportionately young and female to represent the general population.[106]

The 1990 decision in *Moore* did little to quell the sense of uncertainty amongst British researchers. As Brahams again noted in the *Lancet*, and as attendees at a FRAME conference argued, there remained an 'urgent requirement for legal guidelines on the recovery, distribution and use of human tissues, so that surgeons and researchers could be made aware of what was and was not allowed'.[107] Whilst Brahams called on the General Medical Council to draw up recommendations, FRAME members put forward their own proposals, urging doctors to obtain patient consent before researching on tissue.[108] This proposal framed patient consent as a means of safeguarding biomedical research and non-animal methods, claiming it would boost public confidence and ensure a ready supply of tissue. Its authors claimed consent would offer 'reassurance that removal of tissues would be properly conducted', guaranteeing that the ambivalence encountered in the 1987 poll 'would be replaced by willingness to cooperate'.[109]

In 1992, the newly established Nuffield Council on Bioethics announced that it had convened a working party to scrutinize the 'legal and ethical' issues raised by research on human tissues.[110] The Nuffield Council on Bioethics had been created in December 1990, following lobbying by the academic physician Sir David Weatherhall and Stephen Lock, editor of the *British Medical Journal*, who both claimed that a permanent ethics council was vital to maintaining political and public confidence in biomedical research.[111] By promising guidance on the use of human tissue, the Nuffield Council appeared to be fulfilling its remit to educate the public and advise scientists on sensitive issues. The working party was a model of the so-called 'Great and Good' who regularly made up British committees of inquiry. Its chair was the renowned microbiologist Dame Rosalind Hurley; and the seven other members were Ian Kennedy, the Cambridge philosopher Onora O'Neill, the industrialist Trevor Jones, the lawyer Gerald Dworkin, the patent solicitor Kevin Mooney, the pathologist Sir Colin Berry and the writer Kathleen Berry.[112]

When their report was published in 1995, the Nuffield working party claimed it had originated from the 'public concern' surrounding research on human tissue and cells.[113] They argued that *Moore*, which they identified as a '*cause célèbre*', raised several issues that needed urgent attention:

> For example, what relationship exists between the person who was source of the tissue and the tissue removed? Does tissue remain part of the person in any sense, whether symbolically or in some proprietary sense? Does the person retain any right of control over it or is the consent to removal [in operation] to be regarded as implying abandonment of the tissue?[114]

The working party answered these questions by drawing heavily on Richard Titmuss's influential 1970 study, *The Gift Relationship*, which portrayed the free donation of blood as the embodiment of a communitarian welfare state. The Nuffield report framed research on human tissue as a similarly communal endeavour, 'in which researchers and research participants contribute to research that will benefit human health'.[115] The ownership or commercialization of body parts was anathema to this communitarian stance; and the report proposed that patients should be denied property rights in tissue samples, and that hospitals should be prevented from selling tissues to commercial firms.[116]

Like Lori Andrews, the Nuffield working party justified their proposals by evoking public opinion. But in contrast to Andrews, they argued that 'in the general run of things a person from whom a tissue is removed has not the slightest interest in making any claim to it'.[117] The working party did not draw on any survey to justify this claim, but relied instead on the fact that 'the question of a claim over tissue' had not been tested in the British courts.[118] Although their report was commissioned in the wake of a seeming 'public concern', the Nuffield working party clearly felt that this did not extend to research on tissue removed in clinical procedures. They advised that consent to treatment should continue to be seen as an act of abandonment, where tissue attained the status of 'a *res* (a thing)', and argued that consent forms should simply be modified to state that tissues may be used in research.[119]

Despite the uncertainty following *Moore*, and despite calls for new rules and procedures, the Nuffield working party concluded that the acquisition and use of clinical tissues could continue as before. Their proposals received support from a 1996 paper in the *British Medical Journal*, which also claimed that most people had no interest in claiming ownership of tissue samples. But the word 'most' was telling. While the majority of the 384 patients surveyed endorsed research on excised tissues, there was considerable ambiguity when it came to whom they believed tissues belonged to: 27 per cent believed the hospital owned them; another 27 per cent believed they belonged to no-one; 20 per cent believed the recipient laboratory owned them; while 10 per cent believed they remained the property of the patient or their family (the rest did not answer).[120] To the authors, this demonstrated that 'although 90 per cent of the respondents in this study believed that removed tissue belonged to others or to no one, a considerable minority believed that they retained ownership of removed tissue'.[121] To clarify this situation, and like the Nuffield working party, they recommended consent forms should inform patients that excised tissues may be used in research.

But other doctors and researchers continued to endorse more stringent consent protocols. In 1999, an ad-hoc committee from the UK Co-ordinating Committee on Cancer Research (hereafter UKCCCR) issued guidelines that urged oncologists to secure additional consent from patients before creating a tissue culture or cell line. Their guidelines explicitly framed patients as tissue 'donors', who might be entitled to sue should they discover that tissue was used without their per-mission.[122] The Nuffield working party had deliberately avoided the term 'donor' when they discussed patients whose tissue was removed during clinical treatment. As Onora O'Neill stated, this was because 'by definition, donors give tissue that will not otherwise be removed from them'.[123] The UKCCCR committee rejected the abandonment model, including patients and their relatives among 'the long list of people who might lay claim to the ownership of specimens and their deriv-atives'.[124] Notably, the UKCCCR group did not simply believe that these additional consent measures were needed to forestall a British *Moore*. Like the New Labour government that was elected in 1997, and their Conservative predecessors, they believed that additional consent measures were needed to acknowledge that patients were increasingly viewed as empowered 'consumers' who demanded greater influence over their own treatment.[125]

This stance was reiterated in a new edition of Ian Freshney's tissue culture manual (Freshney was also a member of the UKCCCR commit-tee). Freshney urged researchers to obtain consent from the patient or their family before culturing tissue, even if consent had been obtained for a clinical procedure. He warned that without specific consent for research, 'the legal aspects of ownership of the cell lines that might be derived and any future biopharmaceutical exploitation of the cell lines, their genes, and their products becomes exceedingly complex'.[126] Freshney included a sample consent form that researchers or the surgeon should distribute to patients before treatment: informing them tissue may be used in research and asking that they 'give up any claim that you own the tissue or its components, regardless of the use that may be made of it'.[127]

Retained organs and new legislation: Human tissue in the twenty-first century

Despite the efforts of the UKCCCR group, doctors and researchers favoured the abandonment model proposed by the Nuffield working party.[128] But events in 1999 and 2000 ensured that the use of human tissue became even more contentious – forcing another review of the ways in which it was obtained, stored and used. During a public inquiry into high infant mortality rates at Bristol Royal Infirmary, witnesses admitted that many hospitals retained organs that had been removed from children during postmortem examinations, without the express consent of the parents. Although doctors stressed that these organ collections were a vital teaching resource, they became the centre of public scandal. As part of their neo-liberal commitment to active citizens, the New Labour government that was elected in 1997 endorsed the notion of empowered patients; and the non-consented retention of organs appeared completely at odds to their pledge to create a health service 'built around the patient'.[129] Like Labour politicians, journalists portrayed the retention of human material as a closeted and secretive procedure, completely opposed to the wishes of grieving parents.[130] Distaste increased when reporters claimed that large stocks of retained material were never used for research or teaching, and detailed how several hospitals had sold body parts to commercial firms.[131]

While attention often centred on the retention of whole organs, journalists and ethicists also drew attention to the widespread collection of smaller samples, and their retention as pathology slides and paraffin blocks. In public and political debate, commentators often drew no distinction between larger body parts and small tissue samples. Both were portrayed as symbols of medical malpractice; and many tissue samples were returned to families, and re-buried along with whole organs, in the widespread 'amnesty' that took place from 2000 onwards (in Liverpool, where the Alder Hey Children's Hospital amassed the largest collection of organs and tissues, this 'respectful internment project' was not scheduled to end until January 2010). Discussion quickly encompassed the legality and ethics of collecting any tissue – from living and dead bodies, and from adults as well as children. Support for extended consent measures and acknowledgement of property in tissues grew as journalists, lawyers and ethicists argued that tissue's status as *res nullius* was rendered untenable by the emphasis on patient autonomy and the ability to obtain sensitive genetic data from stored tissues.[132] Calls for greater control over body parts were compounded by many families involved in the scandal, who stated that they did not object to the retention of organs per se, but more the fact that they were not asked.[133]

Medical groups subsequently argued that new legislation was needed for the collection of tissue, including material removed in clinical procedures. In 2001, the MRC stated that 'patients should always be informed when material left over following diagnosis or treatment ... might be used for research'.[134] And a joint report by the MRC and the Wellcome Trust, released the same year, also claimed that patient consent was 'a crucial issue ... especially in the current climate'.[135]

In 2001, the Department of Health issued interim guidelines for the collection of human material. These proposals, entitled *Human Bodies, Human Choices*, reflected Labour's enthusiasm for empowered patients: recommending that storing postmortem organs or tissue without the express consent of the next of kin should be made a criminal offence. This represented a distinct shift from the 1961 Human Tissue Act, which argued that doctors need only make 'reasonable enquiry' in order to retain postmortem material.[136] The *Human Choices* guidelines also covered the use of material from living donors, including samples removed during clinical procedures.[137] They notably argued that even when consent had been obtained for clinical treatment, 'specific consent would be required if any secondary uses were proposed of any organs, tissues or body parts'.[138] This was justified because 'people today expect that, in general, human organs and tissue should be removed, retained or used only for purposes for which they have had the opportunity to give their consent'.[139]

The Department of Health guidelines proposed that researchers would only have to obtain consent for the initial creation of a tissue culture or cell line, with additional consent not needed for future projects as long as the tissue culture was anonymized.[140] They also rejected the possibility of donors and families asserting ownership, and 'receiving the equivalent of "royalties"', if tissue cultures proved commercially lucrative. This was again justified on Lockean grounds, with patients denied ownership because they had not 'contributed work and skill to the development of the product'.[141] The guidelines thus framed consent to research as the point where patients were 'asked to give up any property rights (if such exist) in tissue at the time of donation'.[142]

Following these various guidelines, many scientists who used to informally obtain tissue from surgeons now sought permission from patients.[143] Certain hospitals even refused to store tissue cultures unless researchers could prove they had obtained patient consent.[144] But others were unwilling to seek consent from patients – either because they feared a hostile response, or because they viewed patient consent as an unwelcome burden.[145] In 2002, the charity Cancer Research UK argued this reluctance had fostered a shortfall in tissues that was beginning to threaten research. Kathy Pritchard-Jones, a paediatric oncologist, called for the implementation of a 'presumed consent' system, where all tissue was automatically used in research unless a patient or relative raised specific objection. This, she argued, would still give patients the chance to determine how tissue was used, while also removing the 'extra time and paperwork for pathologists and doctors who already have a heavy workload'.[146]

Amidst this uncertainty, the government introduced a new Human Tissue Bill in December 2003. In line with the *Human Choices* guidelines, the Bill made it a criminal offence to store or use tissue without patient consent. This clearly arose as a consequence of a perceived public demand for control over tissues, with ministers asserting that stringent measures were needed to maintain public confidence in biomedical research. However, many funding organizations and scientists strongly opposed the new Bill – and principally its first clause, which stated that researchers should obtain 'appropriate consent' before working on tissues.[147] The neurobiologist Colin Blakemore, recently appointed head of the MRC, claimed that scientists would construe 'appropriate consent' as consent for every specific use of a tissue sample, and would have to re-contact patients or families each time a tissue sample was used in research.[148]

The government removed the 'appropriate consent' rule from a redrafted Human Tissue Bill, which formed the basis for the new Human Tissue Act that was passed in November 2004. This new Act implemented the presumed consent policy for research on tissues removed during clinical procedures, allowing scientists to obtain material without consent unless a patient raised specific objection.[149] It also stated that researchers would have to apply for licences from a new Human Tissue Authority, which regulated biomedical research on human material, as well as its public display in museums and art galleries.[150]

But the 2004 Human Tissue Act did not close the debate on patient consent or ownership. In the 2009 case *Jonathan Yearworth and others v North Bristol NHS Trust*, the British Court of Appeal departed from the work-and-skill principle when it held that the claimants were entitled to damages after a mechanical fault had destroyed the sperm samples they deposited before undergoing chemotherapy. The Lord Chief Justice ruled that the medical work involved in freezing sperm did not mean that property rights inevitably rested with the hospital or doctors. Instead, he argued that that the patients were entitled to damages because 'by their bodies, they alone generated and ejaculated the sperm'.[151] The Lord Chief Justice hinted that non-reproductive tissues could also be subject to this new ruling, stating that 'developments in medical science now require a re-analysis of the common law's treatment of and approach to the issue of ownership of parts or products of a living human body'.[152]

Shortly after *Yearworth,* the Nuffield Council on Bioethics convened another working party on human tissue, inviting views on whether patients should be granted property rights. At the time of writing, less than six years after the new Human Tissue Act, the legal and ethical status of human tissue is once again the subject of public and ethical debate; and the Nuffield Council may well rule that new guidelines are needed to reflect the *Yearworth* judgement. But these questions are likely to persist whatever the Nuffield Council decides. The factors that made the collection of human tissue problematic, like commercial incentives in research and the emphasis on patient autonomy, remain central to our scientific and political landscape. Questions of consent and ownership are therefore likely to prompt discussion between scientists, ethicists, lawyers and politicians for the foreseeable future.

Conclusions

Susan Lawrence has argued that the *Moore* case represented the 'beginning of a major transformation in Western culture of heretofore "worthless" parts of the self and others'.[153] But this chapter has shown that scientists, lawyers and ethicists had begun to reshape the value of 'worthless' tissues well before *Moore*. The questions regarding consent, disclosure and property that have become synonymous with *Moore* were, in fact, first raised in American debates surrounding foetal tissue culture in the 1970s. These questions did not emerge in Britain during the same period, partly due to the absence of a strong pro-life movement. But debate regarding consent and property did emerge here during the 1980s, as growing numbers of scientists and lawyers claimed that the heightened emphasis on patient autonomy increased the chances of a British *Moore*.

In Britain and the United States, support for consent and ownership often came from academic lawyers and bioethicists. But, crucially, it also came from doctors and scientists. Leonard Hayflick was the first person to assert in court that a patient or family might have property rights in cultured tissues; and during the 1990s, scientists in the British UKCCCR advocated more radical measures than the lawyers and philosophers on the Nuffield Council on Bioethics. These are important points to bear in mind, for contemporary discussion of research on human tissue often presents calls for consent and ownership as emanating from outside biomedicine, establishing a dichotomy that does not stand historical analysis.[154] Detailing how views on consent or ownership traversed law, medicine and bioethics illustrates how these social worlds are, in fact, interdependent and co-constitutive of each other.

All of these arguments for and against extended consent or ownership were based on projections of 'public opinion'. Yet when efforts were actually made to survey this opinion, a complex and ambiguous picture emerged. How various individuals and groups interpreted this complexity, and sought to represent the 'public' interest, depended on the regulatory model they were endorsing.[155] Lori Andrews refigured the 20 per cent of respondents who claimed to be uneasy about use of body parts as evidence that 'people' cared about what happened to excised tissues. The Nuffield Council, meanwhile, argued that 'in the general run of things' patients did not care what happened to tissue. And in endorsing a presumed consent policy, Kathy Pritchard-Jones claimed 'most people were happy to donate samples if the issue was properly explained'.[156]

These seemingly divergent proposals are also similar in another crucial respect. All shared a reluctance to question whether tissue culture undermined the distinctions on which common law, scientific practices and patent regulation are based.[157] The lack of consensus encountered in this chapter shows that tissue cultures confound legal and social principles of categorization. They could not be unproblematically made to fit legal conceptions of people nor things; and their proposed circulation in libertarian or communitarian models was no less contentious. With this in mind, then, perhaps the greater task for scientists, lawyers, and ethicists does not lie in forcing tools such as tissue culture into established legal categories. Perhaps it lies in asking how they might redefine them.

7
Epilogue: Tissues in Culture

Tissue culture continued to underpin high-profile research and generate headlines as it approached its centenary in 2007. In a front-page story in April 2006, the *Independent* reported that researchers at the United States Institute for Regenerative Medicine, in North Carolina, had transplanted bladders grown in the laboratory into seven patients with poor bladder function. For this 'sensational breakthrough', researchers took cells from the patients' bladders, cultured them *in vitro* and then moulded them around a biodegradable scaffold.[1] Follow-up appointments indicated that the bladders were functioning normally 46 months after the transplant operation.[2] As the *Independent* noted, this represented an important advance for the nascent field of 'tissue engineering'. While cartilage and blood vessels had been grown on organic scaffolds during the 1990s, no-one had yet grown or successfully introduced a whole organ into patients.[3] The editor of the *Lancet* told the BBC that the cultured bladder heralded 'a revolution in transplantation' – making it possible for patients to have body parts replaced by tailor-made organs grown from their own cells.[4] Many commentators claimed the procedure, known as autologous transplantation, would overcome the problems associated with conventional transplants: including rejection of organs from other patients, the allocation of organs and the removal of material from brain dead patients. Biologists and ethicists also claimed that autologous transplants bypassed the need to acquire stem cells from experimental embryos – resolving the most potent criticism of regenerative medicine.[5]

In newspapers and medical journals, these cultured bladders came to embody the prospect of a laboratory based 'spare parts' medicine.[6] But the laboratory was not the only site in which tissue engineering took place. Enthused by tissue engineering's similarity to sculpture methods, and the possibility of offering a commentary on modern biomedicine, the Australian artists Oron Catts and Ionat Zurr founded the Tissue Culture and Art Project in 1996.[7] After learning tissue culture and tissue engineering methods, Catts and Zurr sculpted cells from a mouse cell line onto biodegradable scaffolds shaped like Guatemalan worry dolls, which stood roughly 10mm tall. They cultured these dolls for three weeks, until cells covered the scaffold, and then transferred the Petri dishes into a sealed incubator, or 'bio-reactor', for display at art galleries. To Catts and Zurr, the dolls represented the mixture of awe and unease people felt toward biotechnology, and they encouraged gallery visitors to record their anxieties onto a nearby computer.[8] Their 'semi-living sculptures' offered a commentary on how biology blurred 'the boundaries between what is born and what is manufactured, what is animate and what is inanimate, and further challenges our perceptions and our relationships with our bodies and our constructed environment'.[9]

In April 2000, Catts and Zurr established their own laboratory, SymbioticA, in the School of Anatomy and Human Biology at the University of Western Australia. Here, they encouraged visiting artists and scientists to use utilize tissue culture and other biological methods as 'a new artistic palette'.[10] They promoted SymbioticA as an 'ambiguous' interdisciplinary

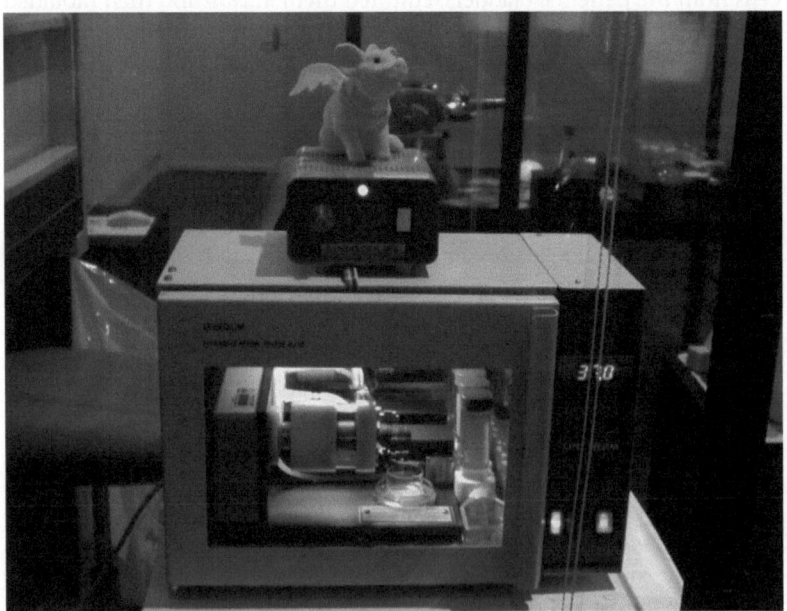

Figure 7.1   Photo from the *Pigs Might Fly* installation, showing the incubator containing 'pig-wing' cultures. Courtesy of the Tissue Culture and Art Project.

space, where laboratory methods served aesthetic and practical ends. Teaming up with the digital artist Guy Ben-Ary, Catts and Zurr's next project again used tissue engineering methods. To convey skepticism toward the Human Genome Project, which they viewed as deterministic and

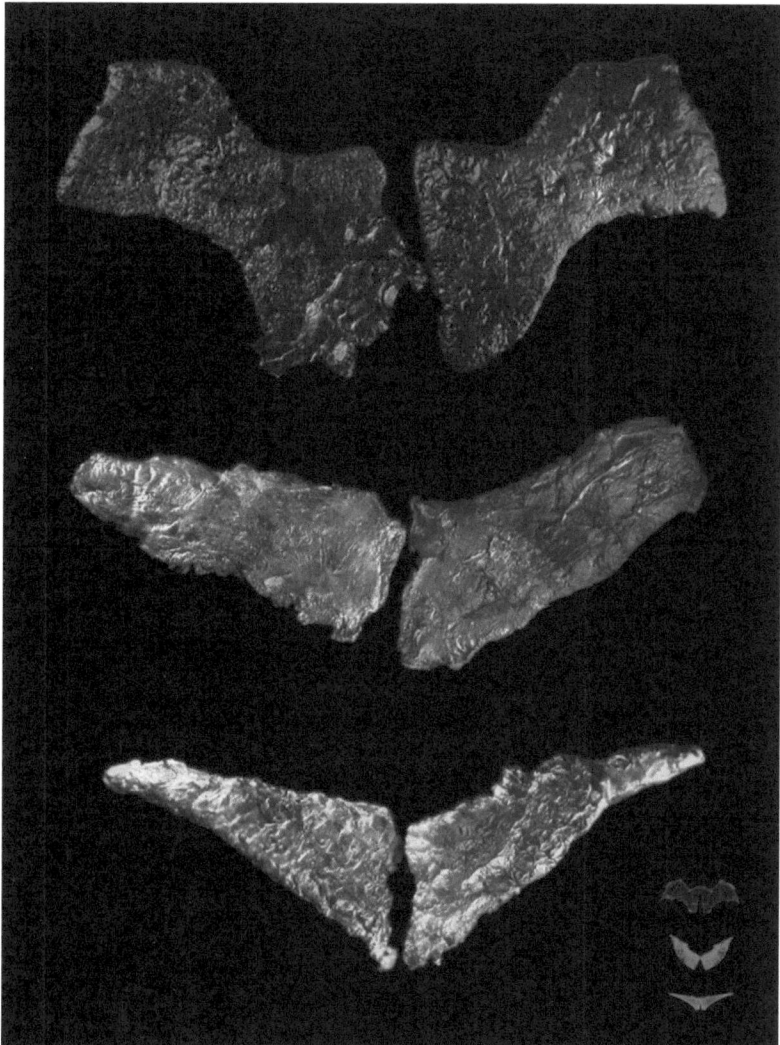

Figure 7.2 Photo from the *Pigs Might Fly* installation, showing the enlarged images of the 'pig wings' that were displayed in art galleries. Courtesy of the Tissue Culture and Art Project.

overblown, they moulded pig stem cells around wing-shaped scaffolds and exhibited the cultures under the title *Pigs Might Fly*. They displayed these 'pig wings' in an incubator, positioned alongside fluorescent images taken under high-magnification (see Figure 7.1 and Figure 7.2). Catts and Zurr also encouraged visitors to watch their daily replenishment of the culture medium: an act that was both essential maintenance and performance art, emphasizing that their exhibit contained 'semi-living entities'.[11]

The Tissue Culture and Art Project has now exhibited its semi-living sculptures in galleries across Australia, Europe and the United States. After the 'pig-wings' project, Catts and Zurr linked their work to animal welfare issues. Their 2003 work *Disembodied Cuisine* displayed cultured frog cells that were seeded onto a scaffold in the shape of a steak. To emphasize how these cultured 'steaks' removed the need for animal slaughter, Catts and Zurr exhibited them alongside the living frog that provided the cells. The installation culminated in a 'feast' at a French gallery, where visitors were invited to eat the cultured steaks. More recently, Catts and Zurr have focussed on the use of animals in the clothing industry. Their *Victimless Leather* installation, exhibited in the New York Museum of Modern Art, and as part of Liverpool's 2008 European City of Culture festival, displayed a culture of mouse cells growing on a scaffold shaped like a miniature jacket.

Like other works in the Tissue Culture and Art Project, and the burgeoning field of 'bioart' more generally, *Victimless Leather* showed how biological techniques and objects straddle the boundary between science and art.[12] Creating this installation required considerable skill in laboratory techniques – from the artists who cultured and manipulated tissues, and the biologists who designed the gallery 'bioreactor'. What is more, while Catts and Zurr considered *Victimless Leather* to be an artistic comment on the exploitation of non-human animals, their scientific collaborators framed it as a practical demonstration of 'the principles and techniques which could underpin future medical therapies'.[13] Like the cultured bladders, it was simultaneously a scientific and a cultural artefact: facilitating new relations between professions, resonating with different groups and linking technical considerations to broader concerns.[14]

To Catts and Zurr, tissue cultures have only recently acquired these multiple and complex meanings. Writing on the technique's history, they state that: 'Initially, the existence of the semi-living, a part of a complex living being sustained alive outside and independent from that being, was rarely discussed, mainly due to its confinement to a scientific context'.[15] They claim it is only since tissue cultures have been wrested from the lab 'and into an artistic context' that we have opened 'discourses about the new relationships we might form with these entities', and begun to re-evaluate our perceptions of the body, life, death and species.

I have shown, on the contrary, that tissue culture resonated beyond the laboratory since its development in 1907. While tissues may not have literally grown outside laboratories until the late 1990s, they appeared in popular sources throughout the twentieth century – in novels, newspapers, science-fiction magazines, on cinema screens and on television. What is more, these popular representations drew upon, and influenced, scientific practices. From the outset, then, tissue culture bound laboratory practices to their broader socio-cultural milieu. It was given meaning by, and helped constitute, a dynamic network of biologists, novelists, journalists and film makers, which later incorporated ethicists, lawyers, anti-vivisection campaigners and pro-life groups.[16] The interplay between these groups ensured that tissue cultures were culturally visible objects long before the era of tissue engineering and bio-art.

Like dissected corpses in the early nineteenth century and vivisected dogs in the late nineteenth century, tissue cultures have longed functioned as publicly iconic objects.[17] They were central to the popular discussion of medicine and biological science throughout the twentieth century, embodying the aspirations, benefits and pitfalls of a highly manipulative, laboratory-based approach to the study of life. Many of the underlying themes in this public interest had a long history, and were raised in previous discussion of research on corpses and experimental animals – such as the boundary between life and death, the natural and artificial, and the tension between interfering with or respecting natural processes.[18] But this does not mean that concerns surrounding tissue culture remained the same over time. Each chapter in this book has demonstrated how certain tissue cultures became objects of interest, or unease, for different and historically contingent reasons, and how debates on broad issues like the scientific manipulation of life, death and the body reflect specific socio-cultural circumstances. For example, during the 1920s and 1930s, Alexis Carrel's old strain, Thomas Strangeways's immortal sausage and Honor Fell's embryo cultures reflected the modernist reconfiguration of lifespan, the body and reproduction. After World War II, discussion of tissue culture's role in vaccine production conveyed enthusiasm for medical progress and 'magic bullets'. Popular attitudes and scientific practices had shifted again by the 1960s, when the creation of hybrid cells contributed to skepticism toward the 'biological revolution'. From the 1970s, the scientific and public struggles over WI-38 and Mo reflected the emergence of pro-life politics, questions over commercial incentives in biomedicine and the increasing emphasis on patient autonomy.

This evidence challenges assumptions that are often made regarding the history of research on tissues. As outlined in the introduction, other commentators have recognized a long history, but framed it on a simple, constant model. They point to a 'fearful symmetry' between the past and present – for scientific practices and public responses.[19] In this model, scientists have long seen the body as a reducible commodity and have secretively pursued research on tissue. Public resistance stems from the way these practices contravene the popular emphasis on bodily integrity, disclosure and right to self-control. This dichotomy is used to explain all resistance to the biomedical use of body parts – from eighteenth and nineteenth century unrest at grave-robbing and dissection, to contemporary events such as the *Moore* case and the retained organs scandal.

By portraying scientists as shadowy figures who regularly contravene popular norms, this model runs the risk of deepening what several commentators identify as a 'culture of mistrust' towards science and medicine.[20] As Onora O'Neill notes, recent controversies surrounding the removal of tissues, the origins of BSE (commonly known as 'mad cow disease') and the development of genetically modified (GM) crops point to a 'systematic crisis of public trust' in scientists and doctors.[21] But, O'Neill continues, there is good evidence that those who seek to remedy this apparent crisis only make it worse – by presenting doctors and scientists as untrustworthy agents who 'pursue their own interests rather than those of patients or of the public'.[22] If histories of objects like tissue culture can contribute to ongoing debates on the removal, storage and use of human tissues, it is by refuting this false dichotomy between science and the public.

We have seen here that scientists and their popular audiences are not diametrically opposed, but are rather interdependent groups that exchange views on tissue culture. This awareness helps us redefine tissue cultures, like other experimental objects, as *sites of engagement*, where scientific and popular interests converge.[23] The conditions and nature of this engagement are not fixed, but are driven by historically specific factors, including research agendas, financial pressures, cultural concerns and political, legal or ethical standpoints. Controversy surrounding the use of tissue arises when these factors come into conflict; and the progress or resolution of specific controversies can re-shape the terms of engagement and the status of tissue itself. Moreover, views on specific factors, like the merits of tissue culture, 'test-tube babies' and patient consent, consistently differed *within* and overlapped *between* social worlds. Rather than think in terms of 'science' and 'public', therefore, it may be better to acknowledge the ongoing formation of hybrid sciences-and-publics, which reconfigure professional and social groups by incorporating actors, objects, ideas and forms of expertise.[24]

Acknowledging how tissue binds different groups together is pertinent when we consider that the term 'tissue' derives from the Latin *texere*, which means 'to weave'. In the medieval period a 'tissue' was a rich cloth, where different elements were entwined. During the eighteenth century it took on a more abstract meaning and denoted a web of statements, errors or absurdities. 'Tissue' did not gain medical currency until *circa* 1800, when the French anatomist Xavier Bichat applied it to the common structural elements shared by organs: ordering the body by classifying 21 types of tissue, and understanding disease as lesions of these specific 'textures'.[25] Each of these different meanings persists, and co-exists, today. As we have seen, a 'tissue' can simultaneously be understood as manufactured, metaphorical or medical material. With this in mind, the French philosopher Michel Serres has claimed that tissue provides an excellent model for apprehending the way the world is ordered – interwoven with different materials, ideas, people and practices.[26]

This book has similarly outlined how tissue culture provides a good model for apprehending the engagement between science and its wider publics. Its history cautions against adherence to universal and totalizing categories, and shows how emergent objects can bring about new relations and possibilities that change our understanding of the taken-for-granted.[27] Tissue culture confounded distinctions between nature and artifice, life and death, human and animal. In the process, it blurred the boundaries between scientific and popular activities, as researchers who used tissue cultures wrote for popular audiences, made films and established new relations with journalists, authors, social groups, lawyers and ethicists.

Highlighting this historical interplay can help foster Onora O'Neill's goal of a 'culture of solidarity', where scientists, patients and the broader public engage in a sustained dialogue on the removal, storage and biomedical use of tissues.[28] This dialogue is vital to maintaining public trust in an era when human tissues remain central to biomedical research and the so-called 'knowledge economy'. Today, more than ever, governments, funding agencies, scientists and patients invest great hopes in research that depends on tissues or cells – whether in tissue engineering, stem cell research, vaccine development or the construction of 'bio-banks'. Confidence in these projects may falter if we continue to believe that scientists and the public have always been at loggerheads over research on tissues. If history can teach us anything, it is that such research offered a rich source of interplay, and that all sides suffered when this engagement was curtailed. Ethicists and policymakers alike should draw comfort from the insight that tissue culture has long prompted a dialogue that is vital to the future of biology and medicine.

Notes

Chapter 1   Introduction

1 Honor B. Fell to Sir Henry Dale (4 February 1935). Held at the Wellcome Trust Library for the History of Medicine, Archives and Manuscripts (hereafter Wellcome Archives), SA/SRL/C.4.
2 Sir Henry Dale to Honor B. Fell (5 February 1935). Wellcome Archives, SA/SRL/C.4.
3 David Masters, 'Science Gets Its Biggest Thrill from the Spark of Life', *Tit-Bits* (3 December 1932).
4 Fell to Dale (5 February 1935).
5 'Tissue culture' is a blanket term that covers the culture of cells and whole organs, known as 'cell culture' and 'organ culture' respectively.
6 A. McGehee Harvey, 'Johns Hopkins – The Birthplace of Tissue Culture: The Story of Ross G. Harrison, Warren H. Lewis and George O. Gey', *The Johns Hopkins Medical Journal*, Vol. 136 (1975), pp. 142–9, on p. 142. See also Meyer Friedman and Gerald W. Friedland, *Medicine's 10 Greatest Discoveries* (London: Yale University Press, 1998), especially Chapter Seven, 'Ross Harrison and Tissue Culture', pp. 133–52.
7 Hannah Landecker, *Culturing Life: How Cells Became Technologies* (Cambridge, Mass: Harvard University Press, 2007).
8 For examples, see Lori Andrews and Dorothy Nelkin, *Body Bazaar: The Market for Human Tissue in the Biotechnology Age* (New York: Crown Publishers, 2001); Ruth Richardson, *Death, Dissection and the Destitute* (Second Edition: London: Phoenix Press, 2001); idem, 'Fearful Symmetry: Corpses for Anatomy, Organs for Transplant?', in Stuart J. Younger, Renée C. Fox and Laurence J. O'Connell (eds), *Organ Transplantation: Meanings and Realities* (Madison: University of Wisconsin Press, 1996), pp. 66–100; Andrew Kimbrell, *The Human Body Shop: The Cloning Marketing and Engineering of Life* (Washington DC: Regenery Publishers, 1997).
9 Lori Andrews and Dorothy Nelkin, 'Whose Body is it Anyway? Disputes over Body Tissue in a Biotechnology Age', the *Lancet*, Vol. 351 (1998), pp. 53–7, on p. 53.
10 John Pickstone, 'Objects and Objectives: Notes on the Material Cultures of Medicine', in Ghislane Lawrence (ed.), *Technologies of Modern Medicine* (London: Science Museum, 1994), pp. 13–24.

Chapter 2   'Make Dry Bones Live': Tissue Culture at the Cambridge Research Hospital

1 Anon., 'Current Topics: Growth of Tissue', *The Medical Press* (31 May 1933).
2 See Timothy Armstrong, *Modernism, Technology and the Body: A Cultural Study* (Cambridge: Cambridge University Press, 1998); Stephen J. Kern, *The*

Culture of Time and Space, 1880–1918 (Cambridge, Mass: Harvard University Press, 1983).

3 Philip J. Pauly, 'Modernist Practice in American Biology', in Dorothy Ross (ed.), *Modernist Impulses in the Human Sciences, 1870–1930* (Baltimore: Johns Hopkins University Press, 1994), pp. 272–89.

4 Richard Overy, *The Morbid Age: Britain Between the Wars* (London: Allen Lane, 2009).

5 John V. Pickstone, *Ways of Knowing: A New History of Science, Technology and Medicine* (Manchester: Manchester University Press, 2000); see also Gerald Geison, *Michael Foster and the Cambridge School of Physiology: The Scientific Enterprise in Late Victorian Society* (Princeton: Princeton University Press, 1978).

6 Jane M. Oppenheimer, 'Taking Things Apart and Putting Them Back Together Again', *Bulletin of the History of Medicine*, Vol. 52, no. 2 (1978), pp. 149–61.

7 Harrison's experiment has been well documented by historians, as an important episode in the history of embryology and as the origin of tissue culture. See Hannah Landecker, 'New Times for Biology: Ross Harrison and the Development of Cellular Life *In Vitro*', *Studies in the History and Philosophy of Biological and Biomedical Sciences*, Vol. 33C, no. 4 (2002), pp. 667–94 (2002); Jane Maienschein, *Transforming Traditions in American Biology* (Baltimore: Johns Hopkins University Press, 1991); Jan A. Witkowski, 'Ross Harrison and the Experimental Analysis of Nerve Growth: The Origins of Tissue Culture', in T.J. Horder, J.A. Witkowski and C.C. Wylie (eds), *A History of Embryology: The Eighth Symposium of the British Society for Developmental Biology* (Cambridge: Cambridge University Press, 1986), pp. 149–77; Donna Haraway, *Crystals, Fabrics, and Fields: Metaphors that Shape Embryos* (London: Yale University Press, 1976).

8 This was also the case for a broader debate regarding the origins and organization of the whole nervous system. For detail on how fixed specimens lent themselves to varying interpretations, see Lorraine Daston and Peter Galison, *Objectivity* (New York: Zone Books, 2007), on pp. 115–20.

9 Ross Harrison, 'Experimental Biology and Medicine', *Physician and Surgeon*, Vol. 34 (1912), pp. 49–65, on p. 56.

10 Harrison, 'Experimental Biology and Medicine' (1912), p. 55.

11 Landecker (2007), pp. 46–7.

12 See Perrin Secler, 'Standardizing Wounds: Alexis Carrel and the Scientific Management of Life in the First World War', *British Journal for the History of Science*, Vol. 41 (2008), pp. 73–109; Andres Horacio Reggiani, *God's Eugenicist: Alexis Carrel and the Sociobiology of Decline* (New York: Berghan Books, 2007). This work also led to criticism from anti-vivisectionists. See Susan Lederer, *Subjected to Science: Human Experimentation in America Before the Second World War* (Baltimore and London: Johns Hopkins University Press, 1995).

13 Hannah Landecker, 'Building "A New Type of Body in which to Grow a Cell": Tissue Culture at the Rockefeller Institute, 1910–1914', in Darwin Stapledon (ed.), *Creating a Tradition of Biomedical Research: Contributions to the History of the Rockefeller University* (New York: Rockefeller University Press, 2004), pp. 151–74.

14 Michael Worboys, *Spreading Germs: Disease Theories and Medical Practice in Britain, 1865–1900* (Cambridge: Cambridge University Press, 2000).

15 Phillip J. Pauly, *Biologists and the Promise of American Life: From Meriwether Lewis to Alfred Kinsey* (Princeton: Princeton University Press, 2000), p. 8.

16 Jan A. Witkowski, 'Alexis Carrel and the Mysticism of Tissue Culture', *Medical History*, Vol. 23 (1979), pp. 279–6.

17 Alexis Carrel, 'On the Permanent Life of Tissue outside of the Organism', *Journal of Experimental Medicine*, Vol. 15 (1912), pp. 516–28.

18 Carrel, 'On the Permanent Life of Tissues' (1912), p. 516.

19 Carrel (1912), p. 516.

20 For more background, see Jon Turney, *Frankenstein's Footsteps: Science, Genetics and Popular Culture* (New Haven and London: Yale University Press, 1998), pp. 73–7.

21 See, for example, Anon., 'Immortal Body Tissues: Cells' Twenty-First Birthday', The *Observer* (22 January 1933); idem, 'Chicken Heart Dead After 28 Years of "Life"', *The Wisconsin State Journal* (17 January 1940); idem, 'Famed Chicken Heart Dies After Long, Artificial Existence', *The Herald Tribune* (2 October 1946).

22 See Kern, *Culture of Time and Space* (1983).

23 Alexis Carrel, 'Physiological Time', *Science*, Vol. 74 (1931), pp. 618–21.

24 Albert Ebeling, 'A Ten Year Old Strain of Fibroblasts', *Journal of Experimental Medicine*, Vol. 35 (1922), pp. 755–9, on p. 756.

25 Anon., 'Will Mankind Learn to Live Forever?' *The Fort Wayne Gazette* (14 October 1929); idem, 'Science May be Able to Suspend Life, Renew It', *The Raleigh Register* (13 December 1935).

26 Anon., 'The Independent Life of Tissues', the *Lancet*, Vol. 195 (1920), p. 335.

27 Anon., 'The Independent Life of Tissues' (1920).

28 Anon., 'Is This Man Quite Mad?' the *Daily Mirror* (14 August 1939), p. 17.

29 Anon., 'Current Topics: Growth of Tissue' (1933).

30 Anon., 'Cambridge Hospital for Special Diseases', *British Medical Journal* (April 1911), pp. 764–6, on p. 764.

31 Anon., *The Strangeways Research Hospital, formerly Cambridge Research Hospital* (Cambridge: Heffer and Sons, 1929), p. 3. Wellcome Archives, SA/SRL/J.3/1.

32 Anon., 'The Research Hospital at Cambridge', the *Lancet*, Vol. 172 (1908), p. 1698.

33 Dorothy E. Strangeways, '1905–1926', in Dorothy E. Strangeways, Frederick G. Spear and Honor B. Fell, *History of the Strangeways Research Laboratory, 1912–1962* (Cambridge: Heffer and Sons, 1962), pp. 7–12.

34 Anon., 'Thomas Strangeways Pigg Strangeways', the *Lancet*, Vol. 209 (1927), p. 56.

35 David Cantor, 'The Definition of Radiobiology' (University of Lancaster PhD thesis, 1987), p. 271.

36 Mark Weatherhall, *Gentlemen, Scientists and Doctors: Medicine at Cambridge 1800–1920* (Oxford: Boydell Press, 2000), pp. 142–74.

37 John T. Dingle and Honor B. Fell, 'Strangeways Research Laboratory', *The Biologist*, Vol. 31 (1984), pp. 191–7.

38 E.D. Strangeways, '1905–1926' (1962), p. 8.

39 Anon., 'The Research Hospital at Cambridge' (1908).

40 Anon., 'The Cambridge Hospital for Special Diseases', *British Medical Journal* (1910), p. 522.
41 Anon., 'Cambridge Hospital for Special Diseases' (1911), p. 764.
42 Thomas Strangeways, *The New Research Hospital at Cambridge* (Cambridge: Heffer and Sons, 1912), p. 4.
43 Anon., 'Cambridge Hospital for Special Diseases' (1911), p. 765.
44 This included specimens donated by the Survey Department of the Egyptian Ministry of Finance, which included Nubian bones estimated to be 6,000 years old.
45 G.E.H. Foxon, 'The Strangeways Research Laboratory' (1980). This is an unpublished manuscript for a history of the Research Hospital, held at the Wellcome archives, SA/SRL/J.6.
46 Foxon, 'Strangeways Research Laboratory' (1980).
47 E.D. Strangeways (1962), p. 12.
48 T.S.P. Strangeways, 'Observations on the Changes Seen in Living Cells During Growth and Division', *Proceedings of the Royal Society of London: Series B, Containing Papers of a Biological Character*, Vol. 94 (1922), pp. 137–41.
49 T.S.P. Strangeways, *The Technique of Tissue Culture 'In Vitro'* (Cambridge: Heffer and Sons, 1924), pp. 26–30.
50 E.D. Strangeways (1962), p. 12.
51 Anon., *The Strangeways Research Laboratory* (1929), p. 4.
52 On the origins of the Medical Research Committee and the formation of the Medical Research Council, see Joan Austoker and Linda Bryder (eds), *Historical Perspectives on the Role of the MRC* (Oxford: Oxford University Press, 1989).
53 On Fletcher and the experimental ethos in nineteenth century Britain, see Geison, *Michael Foster and the Cambridge School of Physiology* (1978).
54 Joan Austoker, 'Walter Morley Fletcher and the Origins of a Basic Biomedical Research Policy', in Austoker and Bryder (eds), *Historical Perspectives on the Role of the MRC* (1989), pp. 23–35.
55 Cantor, 'The Definition of Radiobiology' (1987), p. 275.
56 T.S.P. Strangeways to Malcolm Donaldson (11 March 1924). Wellcome Archives PP/FGS/C.2.
57 Linda Bryder, 'Tuberculosis and the MRC', in Austoker and Bryder (1989), pp. 3–23, on pp. 4–5.
58 T.S.P. Strangeways and H.E.H. Oakley, 'The Immediate Changes Observed in Tissue Cells After Exposure to Soft X-Rays while Growing in Vitro', *Proceedings of the Royal Society of London. Series B, Containing Papers of a Biological Character*, Vol. 95 (1923), pp. 373–81.
59 T.S.P. Strangeways and H.B. Fell, 'Experimental studies on the differentiation of embryonic tissues growing *in vivo* and *in vitro*', *Proceedings of the Royal Society of London. Series B, Containing Papers of a Biological Character*, 99 (1926), pp. 340–66; H.B. Fell, 'The development *in vitro* of the isolated otocyst of the embryonic fowl', *Archiv für Experimentelle Zellforschung*, Vol. 7 (1928), pp. 69–81.
60 Honor B. Fell, 'Cell Biology', in Strangeways, Spear and Fell, *History of the Strangeways Research Laboratory* (1962), pp. 19–33.
61 This forms part of Honor Fell's collection of photos from the 1920s and 1930s. Wellcome Archives, PP/HBF/F.15.

62 Anon., 'Tissue Culture', the *Lancet*, Vol. 201 (1923), p. 858.

63 Anon., 'Recent Developments in Tissue Culture', the *Lancet*, Vol. 204 (1924), pp. 72–3, on p. 73.

64 Anon., 'Tissue Culture', *British Medical Journal* (1924), pp. 152–4, on p. 152. For public coverage of this speech, see Anon., 'The Immortal Cell: Dr Carrel on Tissue Culture', *The Times* (24 July 1924).

65 D. Thomson, 'Controlled Growth en masse (Somatic Growth) of Embryonic Chicken Tissue in Vitro', *Proceedings of the Royal Society of Medicine: Laboratory Reports*, Vol. 7 (1913), p. 77; D. Thomson and J.G. Thomson, 'The Cultivation of Human Tumour Tissue in Vitro', *Proceedings of the Royal Society of Medicine, London*, Vol. 7, no. 1 (1913–14), pp. 7–20; idem, 'The Cultivation of Human Tumour Tissue in Vitro – Preliminary Note', *Proceedings of the Royal Society of Science: Series B, Biology*, Vol. 88 (1914), pp. 90–1.

66 Albert J. Walton, 'The Effect of Various Tissue Extracts Upon the Growth of Adult Mammalian Cells in Vitro', *Journal of Experimental Medicine*, Vol. 20 (1914), pp. 554–72.

67 See A.H. Drew, 'Three Lectures on the Cultivation of Tissues and Tumours in Vitro', the *Lancet*, Vol. 201 (1923), pp. 834–5; E.N. Willmer, 'Studies on the Influence of the Surrounding Medium on the Activity of Cells in Tissue Culture', *British Journal of Experimental Biology*, Vol. 4 (1927), p. 280; idem, 'Tissue Culture from the Standpoint of General Physiology', *Biological Reviews of the Cambridge Philosophical Society*, Vol. 3, no. 4 (1928), pp. 271–302.

68 Willmer, 'Tissue Culture from the Standpoint of General Physiology' (1928), p. 271.

69 F.G. Spear, transcript for 'Tissue Culture' lecture (1928) Held at the Wellcome Archives, Frederick Gordon Spear papers, PP/FGS/E.6/4. See also Carrel, 'Physiological Time' (1929), p. 619.

70 Spear, 'Tissue Culture' (1928).

71 Ibid.

72 Strangeways, 'Observations on the Changes Seen in Living Cells' (1922), p. 139.

73 T.S.P. Strangeways, *Tissue Culture in Relation to Growth and Differentiation* (Cambridge: Heffer and Sons, 1924), p. 10.

74 T.S.P. Strangeways, 'The Living Cell', *British Medical Journal* (1926), pp. 596–7, on p. 596. This paper was the transcript of a talk Strangeways gave at the BMA's annual meeting in 1926, held at Nottingham.

75 For background, see Daniel J. Nicholson, 'Biological Atomism and Cell Theory', *Studies in the History and Philosophy of Science, Part C: Studies in the History and Philosophy of Biological and Biomedical Sciences*, Vol. 41 (2010), pp. 202–11; Andrew Reynolds, 'The Theory of the Cell State and the Question of Cell Autonomy in Nineteenth and Early Twentieth Century Biology', *Science in Context*, Vol. 20 (2007), pp. 71–95; idem, 'The Redoubtable Cell', *Studies in the History and Philosophy of Science, Part C: Studies in the History and Philosophy of Biological and Biomedical Sciences*, Vol. 41 (2010), pp. 194–201.

76 Strangeways, 'The Living Cell' (1926), p. 596. Emphasis added.

77 Julian Huxley, 'Searching for the Elixir of Life', *Century Illustrated Monthly*, Vol. 103 (1922), pp. 621–9, on p. 625; idem, 'Elixir Vitae', in Julian Huxley, *Essays in Popular Science* (London: Chatto and Windus, 1926), pp. 128–30.

78 Strangeways, 'The Living Cell' (1926), p. 596.
79 Anon., 'The Independent Life of Tissues' (1920), p. 33; Huxley, 'Searching for the Elixir of Life' (1922), p. 625.
80 T.S.P. Strangeways, 'Death and Immortality' (1926). This was the sixth and final lecture in Strangeways's 1926 course on 'Tissue Culture'. All lectures are held at the Wellcome Archives, SA/SRL/A.27.
81 G.B.P., 'T.S.P. Strangeways: A Great Biologist', *The Times* (30 December 1926), p. 12.
82 Strangeways, 'Death and Immortality' (1926); see also idem, 'The Living Cell' (1926), p. 596.
83 Strangeways, 'Death and Immortality' (1926).
84 Ibid.
85 Anon., 'T.S.P. Strangeways', *British Medical Journal* (1927), p. 82.
86 Anon., 'Science Finds that Our Bones Die Last of All', the *San Antonio Light* (27 March 1927).
87 Anon., 'Science Finds that Our Bones Die Last of All' (1927).
88 Ibid.
89 Cantor (1987), pp. 282–3.
90 Trustee papers and meeting minutes of the Strangeways Research Laboratory, held at Wellcome Archives, PP/FGS/C.9.
91 For biographies of Honor Fell see J.A. Witkowski, 'Honor Fell', *Trends in Biological Sciences*, Vol. 11 (1986), pp. 486–8; Janet Vaughan, 'Honor Bridget Fell: 22 May 1900–22 April 1986', *Biographical Memoirs of Fellows of the Royal Society*, Vol. 33 (1987), pp. 237–59. On Fell's pioneering role as a woman scientist in interwar Britain, see Lesley A. Hall, 'Chloe, Olivia, Isabel, Letitia, Harriette, Honor, and Many More: Women in Medicine and Biomedical Science, 1914–1945', in Sybil Oldfield (ed.), *This Working-Day World: Women's Lives and Culture(s) in Britain, 1914–45* (London: Taylor and Francis, 1994), pp. 192–202.
92 For more information, see Pnina G. Abir-Am, 'The Assessment of Inter-disciplinary Research in the 1930s: The Rockefeller Foundation and Physico-Chemical Morphology', *Minerva*, Vol. 26 (1988), pp. 153–76.
93 Researchers were never paid directly from lab funds, and were instead supported by a variety of funding bodies. The financial background to the Strangeways laboratory is reviewed in Lesley A. Hall, 'Illustrations from the Wellcome Institute Library: The Strangeways Research Laboratory: Archives in the Contemporary Medical Archives Centre', *Medical History*, Vol. 40 (1996), pp. 231–8.
94 H.B. Fell to M. Donaldson, (23 November 1934). Wellcome Archives, SA/SRL/C.3.
95 Timothy Boon, *Films of Fact: A History of Science in Documentary Films and Television* (London and New York: Wallflower Press, 2008); Gary Werskey, *The Visible College: A Collective Biography of British Scientists and Socialists During the 1930s* (London: Allen Lane, 1978).
96 Fell to Donaldson, (23 November 1934).
97 Honor Fell, 'The Life of a Cell', *The Listener* (22 January 1930), p. 147.
98 Oliver Lodge, 'Solving the Mystery of Life', the *Sunday Express* (9 February 1930), p. 10.
99 Lodge, 'Solving the Mystery of Life' (1930).

100   Ibid.
101   H.G. Wells, Julian Huxley and G.P. Wells, *The Science of Life* (collected edition: London: Cassell and Co., 1938), pp. 17–22, 1008–17. See also, Julian Huxley, 'Man as a Relative Being', in Mary Adams (ed.), *Science in the Changing World* (London: Allen and Unwin, 1932), pp. 110–30; idem, 'Science and Health', *The Listener* (8 November 1933), pp. 706–8.
102   Wells, Huxley and Wells, *The Science of Life* (1938), p. 20.
103   Ibid.
104   Fell, 'The Life of a Cell' (1930).
105   Similar films were made of cultured tissue in the United States and elsewhere in Europe. For discussion of these, see Hannah Landecker, 'Microcinematography and the History of Science and Film', *Isis*, Vol. 97 (2006), pp. 121–32; idem, 'Cellular Features: Microcinematography and Film Theory', *Critical Inquiry*, Vol. 31 (2005), pp. 903–37.
106   Scientific Correspondent, 'Progress of Science: Dr Canti's Films', *The Times* (9 May 1927).
107   Anon., 'Use of Kinema in Cancer Research: Studies of Growing Tissue', the *Manchester Guardian* (27 April 1927).
108   A Pathologist, 'Cinematography and the Microscope', the *Listener* (1 May 1935), pp. 740–1, on p. 741. Although the lecture was given anonymously, its subject matter clearly indicates that Canti was the presenter.
109   A Pathologist 'Cinematography and the Microscope' (1935), p. 741.
110   Anon., 'Cancer Film Shown Before the Duke of York', *The Times* (22 February 1933); Anon., 'How Cancer Grows: Radium's Effect on Cells Filmed', *Daily Telegraph* (22 February 1933); Anon., 'Cancer Cells Filmed: To Be Shown at No. 10', *News Chronicle* (20 February 1933).
111   Anon., 'Cancer Film Shown Before the Duke of York', *The Times* (22 February 1933).
112   This information is taken from notes that accompanied a screening of 'The Cultivation of Living Tissue'. This material is held at the Wellcome Archives, CMAC: SA/SRL/J.4.
113   See Boon, *Films of Fact* (2008).
114   Hannah Fraser, 'A Day with the Scientists Looking at Life: News of Things to Come at Cambridge', *Paris Daily Mail* (27 April 1937).
115   See, for examples, Richard Overy, *The Morbid Age: Britain Between the Wars* (London: Allen Lane, 2009); Bernhard Rieger, *Technology and the Cult of Modernity in Britain and Germany, 1890–1945* (Cambridge: Cambridge University Press, 2005); Anna K. Mayer, '"A Combative Sense of Duty": Englishness and the Scientists', in Christopher Lawrence and Anna K. Mayer (eds), *Regenerating England: Science, Medicine and Culture in Interwar Britain* (Amsterdam: Rodopi Press, 2000), pp. 67–107.
116   See Andrew Tudor, *Monsters and Mad Scientists: A Cultural History of the Horror Movie* (Cambridge, Mass: Basil Blackwell, 1989), pp. 29–31. On pre-War ambivalence toward science, see Philip Blom, *The Vertigo Years: Change and Culture in the West, 1900–1914* (London: Weidenfield and Nicolson, 2008); Kern, *Culture of Time and Space* (1983). On how the War increased this ambivalence, see Turney, *Frankenstein's Footsteps* (1998), pp. 96–7.

117   Aldous Huxley, 'Economists, Scientists and Humanists', in Mary Adams (ed.), *Science in the Changing World* (London: George Allen and Unwin, 1933), pp. 208–23, on p. 209.
118   Fell, 'The Life of a Cell' (1930).
119   E.N. Willmer, *Tissue Culture* (London: Methuen & Co., 1935), pp. xiii–xiv.
120   Julian Huxley, 'Progress, Biological and Other', in *Essays of a Biologist* (London: Pelican Books, 1923), pp. 17–65, on p. 35.
121   A Pathologist (1935), p. 741.
122   Armstrong, *Modernism, Technology and the Body* (1998), p. 3.
123   Julian Huxley, 'The Life Cycle', in Julian Huxley, *Essays in Popular Science* (London: Chatto And Windus, 1926), pp. 75–105, on p. 98. See also Reggiani, *God's Eugenicist* (2008).
124   Julian Huxley, 'The Meaning of Death', in *Essays in Popular Science* (1926), pp. 106–27, on p. 115.
125   Willmer (1928), p. 274.
126   T.S.P. Strangeways, 'Differentiation', part of 1926 series of University lectures on tissue culture. Wellcome Archives, SA/SRL/A27.
127   H.M. Carleton, 'Tissue Culture: A Critical Summary', *British Journal of Experimental Biology*, Vol. 1 (1923), pp. 131–51, on p. 140.
128   Julian Huxley, 'Tissue Growth: An Important Factor in Cancer Research: The British Association Meetings', the *Manchester Guardian* (7 August 1926).
129   Wells, Huxley and Wells, *The Science of Life* (1938), pp. 20–1.
130   Ibid, p. 21.
131   Anon., 'Tissue Culture Film: Cancer Research Development', the *Observer* (20 November 1932), p. 11.
132   Otis Adelbert Kline, 'The Malignant Entity', *Amazing Stories*, Vol. 1 (June 1926), pp. 274–86, on p. 276.
133   Masters, 'Science Gets Its Biggest Thrill from the Spark of Life' (1932).
134   *Lights Out*, 'Chicken Heart' (first broadcast 10 March 1937). Transcript and audio are available online at http://davidszondy.com/Radio. htm.
135   Kate Jackson, *George Newnes and the New Journalism in Britain, 1880–1910: Culture and Profit* (Aldershot: Ashgate, 2001).
136   Norah Burke, 'Could You *Love* a Chemical Baby? For That's What Science Looks Like Producing Next', *Tit-Bits* (16 April 1938).
137   See Overy, *The Morbid Age* (2009); Roger Smith, 'Biology and Values in Interwar Britain: C.S. Sherrington, Julian Huxley and the Vision of Progress', *Past and Present*, Vol. 178 (2003), pp. 210–43.
138   E.N. Wilmer, *Tissue Culture* (London: Methuen & Co., 1935), p. v. By this point, Wilmer had moved from Manchester to the University of Cambridge.
139   Wilmer, *Tissue Culture* (1935), p. v.
140   Wilmer (1935), p. v.
141   Anon., *The Strangeways Research Laboratory* (1929), p. 4.
142   Sheila Jasanoff (ed.), *States of Knowledge: The Co-Production of Science and the Social Order* (London and New York: Routledge, 2004).
143   Jackson, *Newnes and the New Journalism* (2001); Mike Ashley, *The Time Machines: The Story of Science-Fiction Pulp Magazines from the Beginning to 1950* (Liverpool: Liverpool University Press, 2000).

Chapter 3 'Could You *Love* a Chemical Baby?' Organ Culture in Interwar Britain

1 L.M.F. Franks, 'Summary and Future Developments', in Michael Balls and Marjorie Monnickendam (eds), *Organ Culture in Biomedical Research: Festschrift for Dame Honor Fell* (Cambridge: Cambridge University Press, 1976), pp. 549–57.
2 Jacques G. Mulnard, 'The Brussels School of Embryology', *International Journal of Developmental Biology*, Vol. 36 (1992), pp. 17–24, on p. 23.
3 Anthony M. Ludovici, *Lysistrata: Woman's Future and Future Woman* (London: Kegan Paul & Co., 1926), p. 93.
4 Anon., 'Woman Scientist Cultivates Life in Bottles', the *Daily Express* (16 March 1936).
5 Honor Fell to Henry Dale (5 February 1935). Wellcome Archives, SA/SRL/C.4.
6 John E. McWhorter and Allen O. Whipple, 'The Development of the Blastoderm of the Chick *In Vitro*', *Anatomical Record*, Vol. 6 (1912), pp. 121–39.
7 McWhorter and Whipple, 'Development of the Blastoderm of the Chick *In Vitro*' (1912), p. 121.
8 McWhorter and Whipple (1912), p. 124.
9 Harrison (1912), p. 60.
10 On Carrel's influence over tissue culture research in the 1910s and 1920s, see Jan A. Witkowski, 'Alexis Carrel and the Mysticism of Tissue Culture', *Medical History*, Vol. 23 (1979), pp. 279–96.
11 Alexander Maximow, 'Tissue Cultures of Young Mammalian Embryos', *Contributions to Embryology*, Vol. 16 (1925), pp. 49–110, on p. 55.
12 Maximow, 'Tissue Culture of Young Mammalian Embryos' (1925), p. 55.
13 Maximow (1925), p. 49.
14 Ibid.
15 Albert Brachet, 'Recherches sur le determinisme heredetaire de l' oeuf des Mammiferes. Development *in vitro* de jeunes vesicules blastodermiques de Lapin', *Archives de Biologie (Liege)*, Vol. 28 (1913), pp. 447–503.
16 David Thomson, 'Some Further Remarks on the Cultivation of Tissues *in vitro*', *Proceedings of the Royal Society of Medicine*, Vol. 7 (1914), pp. 2–46, on p. 34.
17 David Thomson, 'Controlled Growth en masse (somatic growth) of Embryonic Chicken Tissue In Vitro', *Proceedings of the Royal Society of Medicine: Laboratory Reports*, Vol. 7 (1913), pp. 71–5, on p. 75.
18 Thomson, 'Controlled Growth *en masse*' (1913), p. 73.
19 Ibid.
20 T.S.P. Strangeways and Honor B. Fell, 'Experimental Studies on the Differentiation of Embryonic Tissues Growing *in vivo* and *in vitro* – I. The Development of the Undifferentiated Limb Bud (a) when Subcutaneously Grafted into the Post-Embryonic Chick and (b) when Cultivated *in vitro*', *Proceedings of the Royal Society of London. Series B, Containing Papers of a Biological Character*, Vol. 99 (1926), pp. 340–66, on p. 355.
21 Honor Fell 'The Development of Organ Culture', in Michael Balls and Marjorie Monnickendam (eds), *British Society for Cell Biology Symposium 1: Organ Culture in Biomedical Research: Festschrift for Dame Honor Fell, FRS*

(Cambridge: Cambridge University Press, 1976), pp. 1–13, on p. 3. For the original paper, see T.S.P. Strangeways and Honor B. Fell, 'Experimental Studies on the Differentiation of Embryonic Tissues Growing *in vivo* and *in vitro* – II. The Development of the Isolated Early Embryonic Eye of the Fowl when Cultivated *in vitro*', *Proceedings of the Royal Society of London. Series B, Containing Papers of a Biological Character*, Vol. 100 (1926), pp. 273–83.

22 Strangeways, 'The Living Cell' (1926), p. 526.
23 Honor B. Fell, 'The Development *in vitro* of the isolated otocyst of the embryonic fowl', *Archiv für experimentelle Zellforschung*, Vol. 7 (1928), pp. 69–81.
24 Fell, 'Development of Organ Culture' (1976), p. 3.
25 Honor B. Fell and Robert Robison, 'The Growth, Development and Phosphatase Activity of Embryonic Avian Femora and Limb-Buds Cultivated *In Vitro*', *Biochemical Journal*, Vol. 23 (1929), pp. 767–84.
26 Fell (1976), pp. 4–5.
27 Nick Hopwood, 'Embryology', in Peter J. Bowler and John V. Pickstone (eds), *The Cambridge History of Science, Volume 6: The Modern Biological and Earth Sciences* (Cambridge: Cambridge University Press, 2009), pp. 287–316, on pp. 306–7.
28 Alan Robertson, 'Conrad Hal Waddington. 8 November 1905–26 September 1975', *Biographical Memoirs of Fellows of the Royal Society*, Vol. 23 (1977), pp. 575–622.
29 Conrad H. Waddington, 'Induction by Coagulated Organisers in the Chick Embryo', *Nature*, Vol. 131 (1933), pp. 275–6; idem, 'Experiments on the Development of Chick and Duck Embryos, Cultivated *In Vitro*', *Philosophical Transactions of the Royal Society of London. Series B, Containing Papers of a Biological Nature*, Vol. 221 (1932), pp. 179–230.
30 C.H. Waddington and A.J. Waterman, 'The Development *In Vitro* of Young Rabbit Embryos', *Anatomy*, Vol. 57 (1932), pp. 355–70.
31 The proceedings and schedule for this meeting are held in the Cambridge University Library, Department of Manuscripts and University Archives: Catalogue of the papers and correspondence of Joseph Needham, NCUAS: 54.3.95/J.124. The 'Embryologist's Club' does not appear to have convened again until 1948, when it was officially inaugurated in a meeting at King's College, London. The stated aims of the club, set out at this meeting, were to provide informal discussion of embryological problems, generate opportunities for international links and to compile a record of embryological material available in Britain. The proceedings of this meeting are also held at NCUAS: 54.3.95/J.124.
32 Alan Robertson, 'Conrad Hal Waddington' (1977), p. 590.
33 Abir-Am, 'The Assessment of Interdisciplinary Research in the 1930s' (1988).
34 Fell (1930), p. 147.
35 Julian Huxley, 'Tissue Growth: The British Association Meetings' (1927), p. 8.
36 Masters, 'Science Gets Its Biggest Thrill from the Spark of Life' (1932).
37 Burke, 'Could You Love a Chemical Baby? For That's What Science Looks Like Producing Next', *Tit-Bits* (16 April 1938).

38 J.B.S. Haldane, *Daedalus, Or Science and the Future* (London: Kegan Paul & Co., 1924), p. 1.
39 On the background to *Daedalus* and Haldane's wartime experiences, see Ronald Clark, *J.B.S.: The Life and Work of J.B.S. Haldane* (Oxford: Oxford University Press, 1984). See also Turney (1998), pp. 99–101; Susan Squier, *Babies in Bottles: Twentieth-Century Visions of Reproductive Technology* (New Brunswick: Rutgers University Press, 1994), pp. 66–73.
40 Anon., 'The Age of Miracles', the *Observer* (21 December 1924). An 1926 advert for *Daedalus* in the *Manchester Guardian* indicates that the book quickly went through seven impressions. See also Clark, *J.B.S.* (1984), p. 70.
41 Haldane, *Daedalus* (1924), p. 10.
42 Haldane (1924), p. 64.
43 Ibid, pp. 64–5.
44 Daniel Kevles, *In the Name of Eugenics: Genetics and the Uses of Human Heredity* (Cambridge, Mass: Harvard University Press, 1995), pp. 91–2.
45 Kevles, *In the Name of Eugenics* (1995), pp. 113–28; Werskey, *The Visible College* (1978), pp. 96–7.
46 Kevles notes that the young Haldane 'sympathized for a time with aspects of [mainstream eugenics] particularly its denigration of the lower classes and eagerness to reduce their rate of reproduction'. See ibid, p. 123. In an analysis of *Daedalus*, the molecular biologist David Weatherall also claims that: 'At the time Haldane wrote *Daedalus*, he was an enthusiastic eugenicist'. See David J. Weatherall, '*Daedalus*, Haldane, and Medical Science', in Krishna R. Dronamraju (ed.), *Haldane's Daedalus Revisited* (Oxford: Oxford University Press, 1995), pp. 102–24, on p. 112.
47 Haldane (1924), pp. 66–7. Emphasis added.
48 See Overy, *Morbid Age* (2009), pp. 93–9 for discussion of Marie Stopes and eugenics; see also Kevles (1995), pp. 90–1.
49 Overy (2009), p. 96.
50 Haldane (1924), pp. 65, 68.
51 Ibid, p. 74.
52 Ibid, p. 61.
53 Ibid, p. 63.
54 Anon., 'Review of *Daedalus, or Science and the Future*', *Nature*, Vol. 113 (1924), p. 740.
55 In a 1932 publication on the culture of chick and duck embryos, Conrad Waddington outlined how Strangeways had cultured whole chick embryos, but added that 'only a few experiments were made and the results were never published'. See Waddington, 'Development of Chick and Duck Embryos' (1932), p. 181. Honor Fell also claimed that Strangeways enjoyed 'considerable success' in culturing whole chick embryos. See Fell, 'Cell Biology' (1962), p. 20.
56 T.S.P. Strangeways, 'Lecture 1: Tissue Culture' (December 1926). Wellcome archives, SA/SRL/A.27.
57 Strangeways 'Tissue Culture' (1926).
58 Henry Harris, 'This is Not a Prophecy – It's News About ... Test-Tube Babies!', the *Daily Mirror* (19 May 1937).

59　Eden Paul, *Chronos, or the Future of the Family* (London: Kegan Paul, Trench, Trubner & Co., 1929), p. 51.

60　F.E. Birkenhead, *The World in 2030* (London: Hodder and Stoughton, 1930), p. 165.

61　Birkenhead (1930), p. 15.

62　Ibid, p. 169.

63　Julian Huxley, 'The Tissue Culture King', reprinted in Geoff Cronklin (ed.), *Great Science Fiction by Scientists* (New York: Collins Books, 1970), pp. 348–65, on p. 355.

64　Huxley, 'The Tissue Culture King' (1970), p. 359.

65　J.D. Bernal, *The World, The Flesh and the Devil: An Inquiry into the Future of the Three Enemies of the Rational Soul* (Second Edition: London: Jonathan Cape, 1970), p. 37.

66　Bernal, *The World, the Flesh and the Devil* (1970), p. 32.

67　Bernal (1970), p. 38.

68　Ibid, p. 39.

69　Ibid.

70　Ibid.

71　Ibid, p. 35.

72　See Christine Poggi, 'Dreams of Metallized Flesh: Futurism and the Masculine Body', *Modernism/Modernity*, Vol. 4 (1997), pp. 19–43.

73　Bernal (1970), p. 34.

74　Bertrand Russell, *Icarus, or the Future of Science* (London: Kegan Paul, Trench, Trubner & Co., 1924), p. 2.

75　Anthony Ludovici, cited in Dan Stone, 'Ludovici, Anthony Mario (1882–1971)', *Oxford Dictionary of National Biography* (Oxford: Oxford University Press, online edition, 2009).

76　Anthony M. Ludovici, *Lysistrata, or Woman's Future and Future Woman* (London: Kegan Paul, Trench, Trubner & Co., 1924), p. 76.

77　Ludovici, *Lysistrata* (1924), p. 38.

78　Ludovici (1924), p. 88.

79　Ibid, pp. 92–3.

80　Ibid, p. 89.

81　Ibid, pp. 93–5.

82　Ibid, p. 96.

83　Ibid, pp. 91, 93.

84　Ibid.

85　See Robert Crossley, 'Olaf Stapledon and Idea of Science Fiction', *Modern Fiction Studies*, Vol. 32 (1986), pp. 21–42.

86　Olaf Stapledon, *Last and First Men* (London: Millennium Books, 2004), p. xv.

87　See Robert Crossley, *Olaf Stapledon: Speaking for the Future* (Liverpool: Liverpool University Press, 1994).

88　Stapledon, *Last and First Men* (2004), p. 188.

89　Ibid, pp. 188–9.

90　Ibid, p. 189.

91　Ibid, p. 190.

92　Ibid, p. xv.

93　Ibid, pp. 193–4.

94　See Charlotte Sleigh, 'Plastic Body, Permanent Body: Czech Representations of Corporeality in the early Twentieth Century', *Studies in the History and*

Philosophy of Science, Part C. Studies in the History and Philosophy of the Biological and Biomedical Sciences, Vol. 40 (2009), pp. 241–55.
95 See Ludmilla Jordonova, *Sexual Visions: Images of Gender in Science and Medicine Between the Eighteenth and Twentieth Centuries* (Madison: University of Wisconsin Press, 1989).
96 Biographical information in Martin H. Greenberg (ed.), *Amazing Science Fiction Anthology: The Wonder Years, 1926–1935* (London: TSR UK, 1987), p. 318.
97 Francis Flagg, 'The Machine Man of Ardathia', reprinted in Greenberg (ed.), *Amazing Science Fiction Anthology* (1987), pp. 77–95, on pp. 79, 88.
98 Flagg, 'Machine Man of Ardathia' (1987), p. 88.
99 Flagg (1987), p. 80.
100 Ibid, p. 90.
101 David H. Keller, 'A Biological Experiment', reprinted in David H. Keller, *Tales from Underwood* (Jersey: Spearman Press, 1952), pp. 135–52, on p. 138.
102 Keller, 'A Biological Experiment' (1952), p. 135.
103 Ibid, p. 139.
104 Ibid, p. 140.
105 Turney (1998), p. 115. See also, John Harris, *On Cloning* (London: Routledge, 2004); Judith Arlene Klotzko, *A Clone of Your Own? The Art and Science of Cloning* (Oxford: Oxford University Press, 2004); Francis Fukuyama, *Our Posthuman Future: Consequences of the Biotechnology Revolution* (London: Profile Books, 2002).
106 Aldous Huxley, *Antic Hay* (London: Vintage Books, 2004), p. 49.
107 On Huxley's relationship with J.B.S. Haldane, see Nicholas Murray, *Aldous Huxley: An English Intellectual* (London: Little, Brown, 2002); Clark (1984).
108 Aldous Huxley, 'Economists, Scientists, and Humanists', in Adams (ed.), *Science and the Changing World* (1932), p. 222.
109 Huxley (1932), p. 213.
110 This quote on mass-production is taken from Aldous Huxley, 'To the Puritan All Things are Impure', in Aldous Huxley, *Music at Night & Other Essays* (London: Chatto and Windus, 1931), pp. 173–84, on p. 180.
111 Aldous Huxley, *Brave New World* (London: Flamingo Books, 1994), p. 13.
112 Huxley, *Brave New World* (1994), p. 5.
113 Huxley (1994), pp. 1–15.
114 Ibid, p. 190.
115 Ibid, p. 219.
116 Joseph Needham, 'Biology and Mr. Huxley: review of *Brave New World* by Aldous Huxley', *Scrutiny* (May 1932), pp. 76–9. Cf. Turney (1998), p. 116; Squier, *Babies in Bottles* (1994), p. 147.
117 Needham, 'Biology and Mr. Huxley' (1932), p. 78. Emphasis in original.
118 Needham's notes on Waddington's lecture are held in the Cambridge University Library, Department of Manuscripts and University Archives: Catalogue of the papers and correspondence of Joseph Needham: NCUAS 54.3.95/J.234.
119 Honor Fell to Sir Henry Dale (4 February 1935), SA/SRL/C.4.
120 Fell to Dale (1935).
121 Sir Henry Dale to Honor B. Fell (5 February 1935). Wellcome archives, SA/SRL/C.4.

122 Dale to Fell (1935).
123 Honor B. Fell, 'Tissue Culture: The Advantages and Limitations as a Research Method', *British Journal of Radiology*, Vol. 8 (1935), pp. 27–31, on p. 27.
124 Fell, 'Tissue Culture' (1935), p. 27.
125 Special Correspondent, 'Woman Scientist Cultivates Life in Bottles', *Daily Express* (16 March 1936).
126 Special Correspondent, 'Woman Scientist Cultivates Life in Bottles' (1936).
127 Ibid.
128 Harrison Hardy, 'Any TIN in the Sun?', the *Daily Mirror* (20 March 1937).
129 Burke, 'Could You *Love* a Chemical Baby?' (1938).
130 Ibid.
131 Ibid.
132 Honor B. Fell to Archibald Vivian Hill (30 May 1939), Wellcome Archives, PP/HBF/B.1. For an example of Waddington's popular writing see, Conrad H. Waddington, 'Twenty-Five Years of Biology', *Discovery* (May 1935), pp. 134–7.
133 Fell to Hill (1939).
134 Honor Fell, Lecture to Post-Graduate School of Medicine, 'The Technique of Tissue Culture and its Value in Research' (3 March 1937). Wellcome archives, PP/HBF/E.12/3.
135 Fell to Dale (1935).
136 H.B. Fell to M. Donaldson, (23 November 1934). Wellcome Archives, SA/SRL/C.3.
137 For critical analysis of this issue, see Jackson, *George Newnes* (2001); D.L. LeMahieu, *A Culture for Democracy: Mass Communication and the Cultivated Mind in Britain Between the Wars* (Oxford: Clarendon Press, 1988).
138 Lori Andrews and Dorothy Nelkin, 'Whose Body is it Anyway? Disputes over Body Tissue in a Biotechnology Age', the *Lancet*, Vol. 351 (1998), pp. 53–7, on p. 55.
139 Andrews and Nelkin, 'Whose Body is it Anyway?' (1998), pp. 55, 56.
140 Peter J. Bowler, *Science for All: The Popularization of Science in Early Twentieth-Century Britain* (Chicago and London: University of Chicago Press, 2009).

Chapter 4   Converting Human Material into Tissue Culture, *c.*1910–70

1 John E. Harris, 'Structure and Function in the Living Cell', in Michael Johnson and Michael Abercrombie (eds), *New Biology Five* (London: Penguin, 1948), pp. 26–47, on p. 45.
2 Anon., 'Progress Made in Study of Common Cold: Virus Propagation in Human Tissue', *The Times* (28 July 1954).
3 Nick Hopwood, 'Producing Development: The Anatomy of Human Embryos and the Norms of Wilhelm His', *Bulletin of the History of Medicine*, Vol. 74, no. 1 (2000), pp. 29–79; Adele E. Clark, 'Research Materials and Reproductive Sciences in the United States, 1910–1940', in Gerald Geison (ed.), *Physiology in the American Context, 1850–1940* (Bethesda: Williams and Wilkins, 1987), pp. 323–50. For a later case study, see Warwick Anderson,

The Collectors of Lost Souls: Turning Kuru Scientists into Whitemen (Baltimore: Johns Hopkins University Press, 2008).

4 See Hopwood, 'Producing Development' (2000), pp. 38–9. For a first-hand account, see George W. Corner, *The Seven Ages of a Medical Scientist: An Autobiography* (Pennsylvania: University of Pennsylvania Press, 1981).

5 Alexis Carrel and Montrose Burrows, 'Human Sarcoma Cultivated Outside of the Body', *Journal of the American Medical Association*, Vol. 55 (1910), p. 1732.

6 Carrel and Burrows, 'Human Sarcoma Cultivated Outside of the Body' (1910), p. 1732.

7 Ibid. On Carrel's emphasis on rigorous technical training, see Witkowski, 'Alexis Carrel and the Mysticism of Tissue Culture' (1979).

8 For more on the practices that transform natural materials into experimental entities, see Anderson, *Collectors of Lost Souls* (2008), pp. 133–61; Michael Lynch, 'Sacrifice and the Transformation of the Animal Body into a Scientific Object: Laboratory Culture and Ritual Practice in the Neurosciences', *Social Studies of Science*, Vol. 18 (1988), pp. 265–89.

9 Carrel and Burrows (1910).

10 Ibid.

11 Alexis Carrel and Montrose Burrows, 'Cultivation of Tissues in Vitro and its Technique', *Journal of Experimental Medicine*, Vol. 13 (1911), pp. 387–96; idem, 'Cultivation in Vitro of Malignant Tumours', *Journal of Experimental Medicine*, Vol. 13 (1911), pp. 571–5.

12 Montrose Burrows, 'The Cultivation of Human Cancer Cells in Vitro', *Medical Record*, Vol. 86 (1914), p. 649.

13 Burrows, 'Cultivation of Human Cancer Cells' (1914), p. 649.

14 Burrows (1914).

15 Joseph R. Losee and Albert H. Ebeling, 'The Cultivation of Human Tissue in Vitro', *Journal of Experimental Medicine*, Vol. 19 (1914), pp. 593–602, on p. 602.

16 R.A. Lambert, 'Technique of Cultivating Human Tissues in Vitro', *Journal of Experimental Medicine*, Vol. 24 (1916), pp. 367–72; idem, 'The Comparative Resistance of Bacteria and Human Tissue Cells to Certain Common Antiseptics', *Journal of Experimental Medicine*, Vol. 24 (1916), pp. 683–8.

17 Burrows, 'Cultivation of Human Cancer Cells' (1914), p. 649; Lambert, 'Cultivating Human Tissues' (1917), p. 372.

18 Lambert, 'Comparative Resistance of Bacteria and Human Tissue' (1916), pp. 683–4.

19 As listed in a comprehensive bibliography of tissue culture work compiled in 1953. See Margaret Murray and Gregory Kopech, *Bibliography of the Research in Tissue Culture, 1884–1950: An Index to the Literature of the Living Cell Cultivated In Vitro* (New York: Academic Press, 1953). See also Henry Harris, *The Cells of the Body: A History of Somatic Cell Genetics* (Cold Spring Harbor: Cold Spring Harbor Press, 1995), p. 58.

20 David Thomson and John Gordon Thomson, 'The Cultivation of Human Tumour Tissue *in vitro*', *Proceedings of the Royal Society of Medicine, London*, Vol. 7, no. 1 (1914), pp. 7–20, on p. 8.

21 Bland-Sutton served as President of the Royal College of Surgeons between 1923–25. See Anon., 'Sir John Bland-Sutton', *Nature*, Vol. 139 (1937),

pp. 223–4. Thomson and Thomson, 'Cultivation of Human Tissue'; idem, 'The Cultivation of Human Tissue *in vitro* – Preliminary Note', *Proceedings of the Royal Society of London. Series B, Containing Papers of a Biological Character*, Vol. 88 (1914), pp. 90–1, on p. 90.

22   Thomson and Thomson, 'Cultivation of Human Tumour Tissue' (1914), p. 8.

23   Ibid.

24   Thomson and Thomson (1914), p. 20.

25   These specimens were donated to the Royal College of Surgeons following Strangeways's death in 1926, although the collection was destroyed during a bombing raid in World War II. See E.D. Strangeways, '1905–1926' (1962), p. 10.

26   Willmer, 'Tissue Culture from the Standpoint of General Physiology' (1928), p. 284; Strangeways, *The Technique of Tissue Culture* (1924).

27   Willmer (1928), p. 284.

28   Alfred Glücksmann to David Cantor (15 March 1980), correspondence held at the Wellcome Archives, GC/91/C7.

29   Glücksmann to Cantor (15 March 1980). For publications arising from this research, see Alfred Glücksmann, 'Preliminary Observations on the Quantitative Examination of Human Biopsy Material Taken from Irradiated Carcinoma', *British Journal of Radiology*, Vol. 14 (1941), pp. 187–98; idem, 'Quantitative Histological Analysis of Radiation-Effects in Human Carcinomata', *British Medical Bulletin*, Vol. 4 (1946), pp. 26–30; idem, 'The Influence of Tumour Histology, Duration of Symptoms and Age of Patient on the Radiocurability of Cervix Tumours', *British Journal of Radiology*, Vol. 22 (1949), pp. 90–5.

30   Honor Fell, 'Cell Biology', in Strangeways, Spear and Fell, *History of the Strangeways Research Laboratory* (1962), pp. 19–32, on p. 25.

31   For example, in January 1945 Mowlem wrote to Fell that 'I am hoping to bring you up more some skin'. Dr. Mowlem to H.B.F. Fell (12 January 1945). Wellcome Archives, SA/SRL/G.1.

32   In 1937, a Dr. W.W. Woods wrote to Fell and informed her that he had 'reserved for you a portion of the termination of the stomach. The specimen is in 4% formaldehyde … It seems quite unnecessary for you to send a special messenger for this small specimen, so if you let me know, I will send it by post in a bottle sealed as well as I can'. W.W. Woods to H.B.F. Fell (20 March 1937). Wellcome Archives, SA/SRL/J.2.

33   The project had the dual aim of obtaining information on the mechanisms behind tissue repair whilst also devising improvements in clinical treatment. Fell, 'Cell Biology' (1962). More information on the project is held at the Wellcome Archives, SA/SRL/H.4.

34   H.B. Fell to H.T. Laycock (27 September 1939), Wellcome Archives, SA/SRL/H.4.

35   H.T. Laycock to H.B. Fell (5 October 1939), Wellcome Archives, SA/SRL/H.4.

36   Honor B. Fell, 'Fashion in Cell Biology', Presidential Address to the International Society for Cell Biology, Paris (4 September 1960). Wellcome Archives, PP/HBF/E.17.

37   Fell (1962).

38 Spear, 'Tissue Culture' (1928).
39 See for example, Michael G. Mulinos, 'Cycloserine: An Antibiotic Paradox', *Antibiotics Annual* (1956), pp. 131–5; Howard W. Larsh, Stanley L. Silberg and Agnes Hinton, 'Use of the Tissue Culture Method in Evaluating Fungal Agents', *Antibiotics Annual*, Vol. (1957), pp. 918–22; Charles M. Pomerat, 'Use of Tissue Cultures in Drug Testing Operations', in Maurice B. Visscher (ed.), *Methods in Medical Research: Volume 4* (Year Book Publishers, 1951), p. 211. On the development and impact of antibiotics, see Robert Bud, *Penicillin: Triumph and Tragedy* (Oxford: Oxford University Press, 2007).
40 A. Moscona, O.A. Trowell and E.N. Willmer, 'Methods', in E.N. Willmer (ed.), *Cells and Tissues in Culture: Methods, Biology and Physiology, Volume One* (London: Academic Press, 1965), pp. 19–86.
41 K. Russell, 'Tissue Culture – A Brief Historical Review', *Clio Medica*, Vol. 4 (1969), pp. 110–19; Charity Waymouth, 'Construction and Use of Synthetic Media', in Willmer (ed.), *Cells and Tissues in Culture* (1965), pp. 99–132.
42 Phillip R. White, *Cultivation of Animal and Plant Cells* (New York: Ronald Press, 1962).
43 Witkowski (1979).
44 For more detail on George Gey and the roller tube method, see Landecker (2007), pp. 112–19; A. McGehee Harvey, 'Johns Hopkins – The Birthplace of Tissue Culture: The Story of Ross G. Harrison, Warren H. Lewis and George O. Gey', *The Johns Hopkins Medical Journal*, Vol. 136 (1975), pp. 142–9.
45 Landecker (2007), p. 123.
46 A.E. Feller, John F. Enders and T.H. Weller, 'The Prolonged Coexistence of Vaccinia Virus in High Titre and Living Cells in Roller Tube Cultures of Chick Embryo Tissues', *Journal of Experimental Medicine*, Vol. 72 (1940), pp. 367–88, on p. 385.
47 John F. Enders, Thomas H. Weller and Frederick C. Robbins, 'Cultivation of the Lansing Strain of Poliomyelitis Virus in Cultures of Various Human Embryonic Tissues', *Science*, Vol. 109 (1949), pp. 85–7, on p. 86.
48 Landecker (2007), pp. 122–3.
49 Harris, *The Cells of the Body* (1995), pp. 40–1; John Paul, 'Achievement and Challenge', in Claudio Barigozzi (ed.), *Origin and Natural History of Cell Lines* (New York: Alan R. Liss, 1978), pp. 3–10.
50 Joan H. Fujimara, *Crafting Science: A Sociohistory of the Quest for the Genetics of Cancer* (Cambridge, Mass: Harvard University Press, 1996), pp. 44–6.
51 For more on HeLa, see Landecker (2007), pp. 141–79.
52 Audrey Fjelde, 'Human Tumor Cells in Tissue Culture', *Cancer*, Vol. 8 (1955), pp. 845–51, on p. 845.
53 Landecker (2007), pp. 135–7. See also, Russell W. Brown and James H.M. Henderson, 'The Mass Production and Distribution of HeLa Cells at Tuskegee Institute, 1953–1955', *Journal of the History of Medicine*, Vol. 38 (1983), pp. 415–31.
54 Charles M. Pomerat, 'Use of Tissue Cultures in Drug-testing Operations', in Maurice B. Visscher (ed.), *Methods in Medical Research* (Chicago: Year Book Publishers, 1951), pp. 266–71, on p. 267.
55 W.F. Scherer, Jerome T. Syverton and George O. Gey, 'Studies on the Propagation In Vitro of Poliomyelitis Virus', *Journal of Experimental Medicine*, Vol. 97 (1953), pp. 695–715, on p. 707.

56 Howard W. Larsh, Stanley L. Silberg and Agnes Hinton, 'The Use of the Tissue Culture Method in Evaluating Antifungal Agents', *Antibiotics Annual* (1957), pp. 918–22, on p. 922.

57 Anon., 'Congress for Cell Biology Opens: President on Interest in Tissue Culture', *The Times* (29 August 1957).

58 On positive postwar coverage of science and medicine, see Robert Bud, 'Penicillin and the new Elizabethans', *British Journal for the History of Science*, Vol. 31 (1998), pp. 305–33; Jane Gregory and Steve Miller, *Science in Public: Communication, Culture, and Credibility* (London: Basic Books, 1998), pp. 37–9.

59 Fell, 'Fashion in Cell Biology' (1960).

60 Anon., 'Nobel Prize for Medicine: Virus Research in United States', the *Manchester Guardian* (22 October 1954).

61 Alistair Cooke, 'America and Polio: A Day of Rejoicing', the *Manchester Guardian* (16 April 1955).

62 Anon., '"Most Stringent Tests Known" for Polio Vaccine: Reassurance by Research Scientist', the *Manchester Guardian* (9 March 1956).

63 Alistair Cooke, 'Fresh Discovery by Dr Salk: Offshoot of Polio Work may help Cancer Research', the *Manchester Guardian* (28 December 1957).

64 Anon., 'Human Cells Grown in Laboratory', *The Times* (8 September 1960).

65 Anon., 'Advance in Study of Common Cold: Viruses Propagated in Tissue Culture', *The Times* (29 January 1960); idem, 'Tracking Down the Causes of the Common Cold: Slow but Sure Progress Being Made', *The Times* (18 January 1957); idem, 'Step Forward in Fighting Colds: Virus Under Microscope', the *Manchester Guardian* (10 September 1953).

66 Anon., 'Seeking Methods of Protection from Atomic Radiation: Work of the Medical Research Council', the *Manchester Guardian* (21 July 1955).

67 Anon., 'Tracing Heredity in Virus and Man: Cancer Research', *The Times* (5 September 1959); idem, 'No Cancer in Rats from Tobacco Tar Tests – But Overgrowth in Human Lung', the *Manchester Guardian* (11 July 1956).

68 Anon., 'Hair Grows on a Dish', *Daily Mail* (13 February 1950); idem, 'Miss Hardy Grows Hair on a Plate – in Three Weeks', the *Daily Mirror* (13 February 1950).

69 Undated clipping from the *Wellington Post*, held at the Wellcome Archives, PP/FGS/C.31.

70 John E. Harris, 'Structure and Function in the Living Cell', in M.L. Johnson and Michael Abercrombie (eds), *New Biology, Five* (London: Penguin, 1948), pp. 26–47, on p. 26.

71 Harris, 'Structure and Function in the Living Cell' (1948), p. 45.

72 Science Correspondent, 'The Laboratory of the Living Cell', *The Times* (13 June 1958).

73 Honor B. Fell, 'The Cell as an Individual' (March 1962). Wellcome Archives, PP/HBF/E.17/2.

74 Fell, 'The Cell as an Individual' (1962).

75 Andrew Reynolds, 'The Cell's Journey: From Metaphorical to Literal Factory', *Endeavour*, Vol. 31 (2007), pp. 65–70.

76 Fell, 'The Cell as an Individual' (1962).

77 Harris, 'Structure and Function in the Living Cell' (1948), p. 46.

78   Margaret R. Murray, 'Tissue Culture Procedures in Medical Installations, A: Sources and Handling of Material', in Maurice B. Visscher (ed.), *Methods in Medical Research: Volume 4* (New York: Year Book Publishers, 1951), pp. 211–12, on p. 211.

79   Paul, 'Achievement and Challenge' (1978), pp. 4–5.

80   On the American Type Culture Collection, see Toby Appel, *Shaping Biology: the National Science Foundation and American Biological Research, 1945–1975* (Baltimore: Johns Hopkins University Press, 2000). On changes in the freezing, storage and shipping of cultured material, see Landecker (2007), pp. 153–9.

81   Paul (1978), pp. 3–10.

82   Virginia J. Evans, Naomi M. Hawkins, Benton B. Westfall and Wilton R. Earle, 'Studies on Culture Lines Derived from Mouse Liver Parenchymatous Cells Grown in Long Term Tissue Culture', *Cancer Research*, Vol. 18 (1958), pp. 261–6.

83   Cooke, 'Fresh Discovery by Dr Salk' (1959).

84   On the Food and Drug Administration's policy, see Gretchen Vogel, 'FDA Weighs Using Tumor Cell Lines for Vaccine Development', *Science*, Vol. 285 (1999), pp. 1826–7.

85   E.N. Willmer, *Tissue Culture* (Third edition: London: Methuen & Co., 1964), pp. 57, 61.

86   E.N. Willmer, 'Introduction', in Willmer (ed.), *Cells and Tissues in Culture: Methods, Biology and Physiology, Volume One* (1965), pp. 1–17, on p. 11.

87   Professor L.M. Franks, interview with the author (12 May 2004).

88   See Michael Gold, *A Conspiracy of Cells: One Woman's Immortal Legacy and the Medical Scandal it Caused* (State University of New York Press, 1985).

89   Alfred Glücksmann, 'Cell Deaths in Normal Vertebrate Ontogeny', *Biological Reviews*, Vol. 26 (1951), pp. 56–89. See also idem, 'Mitosis and Degeneration in the Morphogenesis of the Human Foetal Lung *In Vitro*', *Zeitschrift fur Zellfurschung und Mikroskopiche Anatomie*, Vol. 64 (1964), pp. 101–10.

90   Ilse Lasnitzki, 'The Effect of 3–4 Benzpyrene on Human Foetal Lung Grown *In Vitro*', *British Journal of Cancer*, Vol. 10 (1956), pp. 510–16; idem, 'Observations on the Effects of Condensates from Cigarette Smoke on Foetal Lung *In Vitro*', *British Journal of Cancer*, Vol. 12 (1958), pp. 547–52; idem, 'The Effect of a Hydrocarbon-enriched Fraction of Cigarette Smoke Condensate on Human Fetal Lung Grown *in Vitro*', *Cancer Research*, Vol. 28 (1968), pp. 510–16.

91   Margaret Brazier, *Medicine, Patients and the Law* (London: Penguin, 2003).

92   Maurice Pappworth, *Human Guinea Pigs: Experimentation on Man* (London: Routledge and Kegan Paul, 1967), pp. 91–4, 125–6.

93   Susan C. Lawrence, 'Beyond the Grave – The Use and Meaning of Human Body Parts: An Historical Introduction', in Robert Weir (ed.), *Stored Tissue Samples: Legal, Ethical and Public Policy Issues* (Iowa: University of Iowa Press, 1998), pp. 111–42, on p. 122.

94   Lynn M. Morgan, *Icons of Life: A Cultural History of Human Embryos* (University of California Press, 2009); idem, '"Properly Disposed Of": A History of Embryo Disposal and the Changing Claims on Fetal Remains', *Medical Anthropology*, Vol. 21 (2002), pp. 247–74.

95 Susan Lederer, *Flesh and Blood: Organ Transplantation and Blood Trans-fusion in Twentieth Century America* (Oxford: Oxford University Press, 2008), pp. 24–5. Lederer has elsewhere argued that where opposition to these experimental procedures did arise, it was likely to emanate from US anti-vivisectionist groups, who linked the dissection of animals for tissue to the mistreatment of patients. See Lederer, *Subjected to Science* (1995), pp. 77–100.
96 Huxley, 'The Tissue Culture King' (1970), pp. 355–8.
97 A Pathologist, 'Cinematography and the Microscope' (1935), p. 740.
98 Anon., 'Tracing Heredity in Virus and Man' (1959).
99 Anon., 'Vaccine Cultivated in Human Embryo Cells: Work on Common Cold Research', the *Guardian* (26 July 1961).
100 Anon., 'Advance in Study of Common Cold' (1960).
101 Ruth Faden, quoted in Rebecca Skloot, 'Henrietta's Dance', *Johns Hopkins Magazine* (April 2000). Available online at http://www.jhu.edu/jhumag/0400web/01.html.
102 Catherine Waldby and Robert Mitchell, *Tissue Economies: Blood, Organs and Cell Lines in Late Capitalism* (Durham and London: Duke University Press, 2006), pp. 33–4. See also Arjun Appadurai (ed.), *The Social Life of Things: Commodities in Cultural Perspective* (Cambridge: Cambridge University Press, 1986).

Chapter 5   'A Cell is Not an Animal': Negotiating Species Boundaries in the 1960s and 1970s

1 Caroline Walker Bynum, *Metamorphosis and Identity* (New York: Zone Books, 2005); Mary Douglas, *Purity and Danger* (London: Routledge, 2002).
2 Lorraine Daston and Katherine Park, *Wonders and the Order of Nature, 1150–1750* (New York: Zone Books, 1998).
3 Daston and Park, *Wonders and the Order of Nature* (1998), p. 176.
4 See Henry Harris, *Cell Fusion: The Dunham Lectures* (Oxford: Clarendon Press, 1970).
5 J.F. Enders and T.C. Peebles, 'Propagation in Tissue Cultures of Cyto-pathogenic Agents from Patients with Measles', *Proceedings of the Society for Experimental Biology and Medicine*, Vol. 86 (1954), pp. 277–86.
6 Yoshio Okada, 'Analysis of Giant Polynuclear Cell Formation Caused by HVJ Virus from Erlich Ascites Tumour Cells: I. Microscopic Observation of Giant Polynuclear Cell Formation', *Experimental Cell Research*, Vol. 26 (1962), pp. 98–107.
7 For biographical background on Henry (now Sir Henry) Harris, see his auto-biography, *The Balance of Improbabilities: A Scientific Life* (Oxford: Oxford University Press, 1987). For general recollections on the development of cell fusion, see Harris, *The Cells of the Body* (1995). For an historical overview of the scientific impact of cell fusion, see Landecker (2007), pp. 180–219.
8 Henry Harris and John F. Watkins, 'Hybrid Cells Derived from Mouse and Man: Artificial Heterokaryons of Mammalian Cells from Different Species', *Nature*, Vol. 205 (1965), pp. 640–6.
9 Harris and Watkins, 'Hybrid Cells Derived from Mouse and Man' (1965).
10 Ibid, p. 646.
11 From Harris and Watkins (1965), p. 642.

12 Anon., 'Mouse-Man Hybrid Cells Produced', *The Times* (13 February 1965).
13 Bryan Silcock, 'Man-animal Cells Are Bred in Lab', the *Sunday Times* (14 February 1965).
14 Harris, *Balance of Improbabilities* (1987), pp. 192–3.
15 Anon., 'Man-Animal Cells are Bred in Lab', the *Daily Mirror* (15 February 1965).
16 Dominic Sandbrooke, *White Heat: A History of Britain in the Swinging Sixties* (London: Little Brown, 2006).
17 Ayesha Nathoo, *Hearts Exposed: Transplants and the Media in 1960s Britain* (Basingstoke: Palgrave, 2009).
18 Jean Rostand, *Can Man Be Modified?* (London: Secker and Warburg, 1959), p. 96.
19 Turney (1998), p. 143.
20 Robert L. Sinsheimer, quoted in *Towards Tomorrow: Assault on Life* (broadcast BBC One, 7 December 1967).
21 Professor Sir Henry Harris in interview with Gordon Wolstenholme (1986), Oxford Brookes Medical History Video Archive, catalogue number MSVA 012.
22 Henry Harris, 'Hybrid Cells from Mouse and Man', *New Scientist* (18 February 1965), pp. 420–2.
23 Henry Harris, quoted in Gerald McKnight, 'Fused Together ... a Man and a Mouse. And the Next Step Could be Tree Men. Why Scientists Create Monsters', *Tit-Bits* (1 July 1965).
24 Harris, quoted in McKnight, 'Fused Together ... a Man and a Mouse' (1965).
25 Henry Harris, quoted in *Towards Tomorrow: Assault on Life* (BBC One, broadcast 7 December 1967).
26 Edmund Leach, *A Runaway World?* (London: British Broadcasting Corporation, 1968), p. 1.
27 E.H.S. Burhop, 'The British Society for Social Responsibility in Science', *Physics Education*, Vol. 6 (1971), pp. 140–2, on p. 140.
28 Jon Agar, 'What Happened in the Sixties?', *British Journal for the History of Science*, Vol. 41 (2008), pp. 567–90.
29 Bertrand Russell, *Has Man a Future?* (Nottingham: Spokesman Books, 2001), p. 24.
30 Edmund Leach, *A Runaway World* (1968), p. 82.
31 Sandbrooke, *White Heat* (2006).
32 Steven Rose and Hillary Rose, *Science and Society* (London: Penguin Press, 1969), p. xii.
33 Gordon Rattray Taylor, *The Biological Time-Bomb* (London: Book Club Associates, 1968), p. 21.
34 Rattray Taylor, *Biological Time-Bomb* (1968), p. 8, pp. 202–31.
35 Turney (1998), pp. 158–9.
36 Dennis Potter 'Biological Revolution: But What Else?', *The Times* (27 April 1968).
37 Potter, 'Biological Revolution' (1968).
38 Taylor (1968), p. 158.
39 Ibid, p. 169.
40 Ibid, pp. 172–3.
41 Ibid, pp. 214–15.
42 Arthur Koestler, 'Science Out on a Limb', the *Observer* (21 April 1968).

43 Gerald McKnight, 'Breakthrough for the Boffins Trying to Answer the Puzzle: Can Life be Created?' *Tit-Bits* (24 June 1967).
44 McKnight, 'Why Scientists Create Monsters' (1967).
45 The song juxtaposed an upbeat melody to rather pessimistic lyrics. The opening lines, for example, proclaimed that: 'It's good news week/Someone's dropped a Bomb somewhere/Contaminating atmosphere and blackening the sky'.
46 *Radio Times* (21 December 1967).
47 Ibid.
48 Roy Battersby, preview to 'Towards Tomorrow: Assault on Life', *Radio Times* (30 November 1967).
49 Battersby (1967).
50 *Towards Tomorrow: Assault on Life* (broadcast 7 December 1967).
51 Stephanie Harrison, 'Responsibility of Scientists to Society', *The Times* (14 December 1967).
52 Harrison, 'Responsibility of Scientists to Society' (1967).
53 BBC Audience Research Report: 'Towards Tomorrow, Assault on Life' (30 January 1968). BBC Written Archives, Catalogue number VR/67/777.
54 Anon., 'Protest Over TV', *The Times* (15 December 1967).
55 Anon., 'Scientists Object to Broadcast', *The Times* (15 December 1967).
56 Anon., 'Scientists Object to Broadcast' (1965).
57 John F. Watkins, 'Scientists and Society', *The Times* (16 December 1967).
58 Anon., 'Scientists Object to Broadcast' (1967).
59 Watkins, 'Scientists and Society' (1967).
60 See Nathoo, *Hearts Exposed* (2009).
61 Taylor (1968), p. 39.
62 R.G. Edwards, B.D. Bavister and P.C. Steptoe, 'Early Stages of Fertilization *in vitro* of Human Oocytes Matured *in vitro*', *Nature*, Vol. 221 (1969), pp. 632–5. On popular coverage of this work, see Turney (1998), pp. 160–88.
63 See, for example, Robert G. Edwards, 'Aspects of Human Reproduction', in Watson Fuller (ed.), *The Social Impact of Modern Biology* (London: Routledge and Kegan Paul, 1971), pp. 108–22. This is the transcript of a public conference held in London.
64 Rattray Taylor (1968), pp. 37–8. See also McKnight, 'Breakthrough for Boffins' (1967), p. 22.
65 Honor B. Fell, 'The Linacre Lecture, 1969. Cells in Captivity: Past, Present and Future', *Journal of the Women's Medical Federation*, Vol. 52 (1970), pp. 32–48, on p. 46.
66 Fell, 'Cells in Captivity' (1970), p. 47. Emphasis in original.
67 Fell (1970), pp. 47–8.
68 Anon., 'Vivisection Fighters Told to Try Science', the *Guardian* (5 September 1968).
69 Anon., 'Vivisection Fighters Told to Try Science' (1968).
70 Lodge, 'Solving the Mystery of Life' (1930).
71 Ibid.
72 Fell (1930).
73 Fell, 'Tissue Culture: The Advantages and Limitations as a Research Method' (1935), p. 27.
74 Fell (1935), p. 30.

75 Ibid, pp. 29–30.
76 Richard D. French, *Antivivisection and Medical Science in Victorian Society* (Princeton: Princeton University Press, 1975).
77 Robert G.W. Kirk, *Reliable Animals, Responsible Scientists: Constructing Standard Laboratory Animals in Britain c.1919–1972* (University of London PhD thesis, 2006).
78 Kingsley F. Sanders, 'Tissue Cultures as Substitutes for Experimental Animals', *Collected Papers of the Laboratory Animals Bureau*, Vol. 6 (1957), pp. 35–44, on p. 42.
79 Sanders, 'Tissue Cultures as Substitutes for Experimental Animals' (1957), p. 37.
80 Sanders (1957), p. 36.
81 Ibid, p. 41.
82 Ibid.
83 William Russell, 'The Increase of Humanity in Experimentation: Replacement, Reduction and Refinement', *Collected Papers of the Laboratory Animals Bureau*, Vol. 6 (1957), pp. 23–7.
84 William Russell and Rex Burch, *The Principles of Humane Experimental Technique* (London: Methuen & Co., 1959), p. 72.
85 Russell and Burch, *Principles of Humane Experimental Technique* (1959), p. 92.
86 Russell and Burch (1959), p. 79.
87 Ibid, p. 82.
88 Ibid, p. 81.
89 Sir Sydney Littlewood (chair), *Report of the Departmental Committee on Experiments on Animals* (London: Her Majesty's Stationery Office, 1965).
90 Littlewood, *Report of the Departmental Committee on Experiments on Animals* (1965), p. 25.
91 Littlewood (1965), pp. 73–6.
92 *Hansard*, Vol. 924 (1977), p. 34.
93 Janet Fookes, MP, quoted in *Hansard*, Vol. 849 (1973), p. 1584.
94 John Vyvyan, *In Pity and in Anger: A Study of the Use of Animals in Science* (London: Joseph Press, 1969), p. 2.
95 Jon Harris, 'Experiments on Animals', *The Times* (21 January 1971).
96 Richard D. Ryder, *Victims of Science: The Use of Animals in Research* (London: Davis-Poynter, 1975).
97 Peter Singer, *Animal Liberation: Toward an End to Man's Inhumanity to Animals* (London: Cape, 1976).
98 Susan E. Lederer, 'Experimentation and Ethics', in Peter J. Bowler and John V. Pickstone (eds), *The Cambridge History of Science, Volume Six: The Modern Biological and Earth Sciences* (Cambridge University Press, 2009), pp. 583–600, on p. 598.
99 Anon., 'Concern Grows on Experiments', the *Guardian* (1 August 1972).
100 Anon., 'Vivisection Fighters Told to Try Science' (1968).
101 Ibid.
102 Editorial, *New Scientist*, Vol. 45 (26 February 1970), p. 424.
103 C.E. Foister, 'Experiments on Animals', *The Times* (29 January 1971).
104 FRAME, *Is the Experimental Animal Obsolete?* (London: FRAME, 1970), p. 1.

105   Air Chief Marshal Lord Dowding, 'Foreword', in John Vyvyan, *In Pity and in Anger* (1969), pp. 7–9.
106   The Trust took its name from Walter Hadwen, a nineteenth century doctor who had served as President of the British Union for the Abolition of Vivisection.
107   Anon., 'Fund to Protect Animals', the *Guardian* (16 March 1973), p. 26.
108   Bernard Conyers, 'Animal Experiments', the *New Scientist* (19 September 1974), p. 757.
109   Anon., 'Help Us Build: The Dr Hadwen Institute for Humane Research', the *Guardian* (7 December 1970), p. 16.
110   Ryder, *Victims of Science* (1975), p. 114.
111   Dowding, 'Foreword' (1969), p. 7.
112   Anon., 'Fund to Protect Animals' (1973).
113   P.J. Kavanagh, 'Animal Harm', the *Guardian* (26 February 1975).
114   Elystan Morgan, *Hansard*, Vol. 849 (1973), pp. 1584–5.
115   Prime Minister James Callaghan, *Hansard*, Vol. 940 (1977), p. 1644.
116   Ena Kendall, 'A Fairer Deal for Animals', the *Guardian* (16 April 1978). See also Pearce Wright, 'Tissue Culture Tests Saving Use of Laboratory Animals', *The Times* (12 April 1978).
117   Bernard Conyers, 'Animal Experiments' (1974).
118   Vaughan, 'Dame Honor Fell' (1987), p. 250.
119   Charles Foister to Honor B. Fell (16 February 1970) Wellcome Archives, SA/SRL/G.52.
120   Honor B. Fell to Charles Foister (4 March 1970) Wellcome Archives, SA/SRL/G.52.
121   Bernard Conyers to Honor B. Fell (25 May 1972) Wellcome Archives, PP/HBF/D.19.
122   Reprinted as Honor B. Fell, 'Tissue Culture and Its Contribution to Biology and Medicine', *Journal of Experimental Biology*, Vol. 57 (1972), pp. 1–13, on p. 11.
123   Fell, 'Tissue Culture and Its Contribution to Science and Medicine' (1972), p. 11.
124   Conyers, 'Animal Experiments' (1974).
125   Fell, 'Development of Organ Culture' (1976), p. 12.
126   Frank Hooley, MP, *Hansard*, Vol. 940 (1977), pp. 1643–4. Hooley's statement was part of the broader discussion in which Prime Minister James Callaghan endorsed tissue culture.
127   Kendall, 'Fairer Deal for Animals' (1978), p. 10.
128   Smyth (1978), p. 166.
129   Smyth (1978), p. 113.
130   Ibid, p. 157.
131   Ibid, p. 114.
132   Ibid, p. 11.
133   Smyth (1978), p. 117.
134   Kendal (1978).
135   For example, Nathoo (2009); Sandbrooke (2006); Turney (1998).
136   Landecker, 'Building "A New Type of Body in which to Grow a Cell"' (2007), pp. 167–8.

Chapter 6 Nobody's Thing? Consent, Ownership, and the Politics of Tissue Culture

1 Helen Busby, 'Informed Consent: The Contradictory Ethical Safeguards in Pharmacogenetics', in Richard Tutton and Oonagh Corrigan (eds), *Genetic Databases: Socio-ethical Issues in the Collection and Use of DNA* (London: Routledge, 2004), pp. 78–97.
2 See Mitchell Dean, *Governmentality: Power and Rule in Modern Society* (London: Sage Publications, 2010); Nikolas Rose, *Powers of Freedom: Reframing Political Thought* (Cambridge: Cambridge University Press, 1999).
3 Henry Stanhope, 'Live Foetuses Sold for Research – MP', *The Times* (16 May 1970). On the 1967 Abortion Act, see Barbara Brooks, *Abortion in England, 1900–1967* (London: Croom Helm, 1988); Naomi Pfeffer and Julie Kent, 'Framing Women, Framing Fetuses: How Britain Regulates Arrangements for the Collection and Use of Aborted Fetuses in Stem Cell Research and Therapies', *Biosocieties*, Vol. 2 (2007), pp. 429–47.
4 Brookes, *Abortion in England* (1988).
5 Anon., 'Use of Live Foetus Backed' *The Times* (18 May 1970).
6 Anon., 'Unborn Babies: Doctors May Get New Code of Practice', the *Daily Express* (19 May 1970). Emphasis in original.
7 Department of Health and Social Security, Scottish Home and Health Department, Welsh Office, *The Use of Foetuses and Foetal Material for Research: Report of the Advisory Group* (London: Her Majesty's Stationary Office, 1972), p. 3.
8 *The Use of Foetuses and Foetal Material for Research: Report of the Advisory Group* (1972), p. 3.
9 Ibid, pp. 8–9.
10 Ibid, p. 12.
11 Ibid.
12 Pfeffer and Kent, 'Framing Women, Framing the Fetus' (2007), p. 433.
13 Lawrence, 'Beyond the Grave' (1998), p. 111.
14 *Doodeward v Spence* (1908), CLR 406, 414.
15 John Locke, *Two Treatises of Government* (Cambridge: Cambridge University Press, 1988), p. 288. For more on the 'work-and-skill' principle, see Loane Skene, 'Who Owns Your Body? Legal Issues in the Ownership of Bodily Material', *Trends in Molecular Medicine*, Vol. 8 (2008), pp. 48–9; Bronwyn Parry and Cathy Gere, 'Contested Bodies: Property Models and the Commodification of Human Biological Artefacts', *Science as Culture*, Vol. 15 (2006), pp. 139–58.
16 L.M.F. Franks, interview with the author (12 May 2004).
17 James Risen and Judy L. Thomas, *The Wrath of Angels: The American Abortion War* (New York: Basics Books, 1998); David J. Garrow, *Liberty and Sexuality: The Right to Privacy and the Making of Roe v Wade* (Oxford: Maxwell Macmillan International, 1994).
18 Adam Hedgecoe, 'A Form of Practical Machinery: The Origins of Research Ethics Committees in the UK, 1967–1972', *Medical History*, Vol. 53 (2009), pp. 331–50, on p. 338. See also Richard Ashcroft and Mary Dixon-Woods,

'Regulation and the Social Licence for Medical Research', *Medical Health Care and Philosophy*, Vol. 11 (2008), pp. 381–91.

19   Department of Health and Social Security (1972), p. 10. Emphasis added.

20   On the emergence of 'bioethics' in the United States, see Renee Fox and Judith Swazey, *Observing Bioethics* (Oxford: Oxford University Press, 2008); M.L. Tina Stevens, *Bioethics in America: Origins and Cultural Politics* (Baltimore: Johns Hopkins University Press, 2003); David Rothman, *Strangers at the Bedside: A History of How Law and Bioethics Transformed Medical Decision Making* (New York: Basic Books, 1991). On the infamous Tuskegee syphilis experiments, see James H. Jones, *Bad Blood* (New York: Free Press, 1981).

21   Paul Ramsey, *The Ethics of Fetal Research* (New Haven: Yale University Press, 1975), p. 67.

22   Ramsey, *Ethics of Fetal Research* (1975), pp. xii, 67.

23   Diana S. Hart, 'Fetal Research and Anti-Abortion Politics: Holding Science Hostage', *Family Planning Perspectives*, Vol. 7 (1975), pp. 72–82, on p. 73.

24   Daniel Callahan, *Abortion: Law, Choice and Morality* (London: Collier-Macmillan, 1972).

25   Leonard Hayflick, Stanley A. Plotkin, Thomas W. Norton and Hilary Koprowski, 'Preparation of Poliovirus in a Human Fetal Diploid Cell Strain', *American Journal of Hygiene*, Vol. 75 (1962), pp. 240–58; Hilary Koprowski, 'Live Poliomyelitis Vaccines: Present Status and Problems for the Future', *Journal of the American Medical Association*, Vol. 178 (1961), pp. 1151–5.

26   Donald J. Merchant (ed.), *Cell Cultures for Virus Vaccine Production* (Maryland: National Institute of Health, 1968).

27   J.P. Jacobs, C.M. Jones and J.P. Baille, 'The Characteristics of a Human Diploid Cell Designated MRC-5', *Nature*, Vol. 227 (1970), pp. 168–70.

28   Leonard Hayflick, 'The Coming of Age of WI-38', in Karl Maramorsch (ed.), *Advances in Cell Culture, Volume 3* (Orlando: Academic Press, 1984), pp. 303–16.

29   Hayflick, 'Coming of Age of WI-38' (1984), p. 313.

30   See Hayflick (1984), p. 303. This paper on cell aging was eventually accepted and published as Leonard Hayflick, 'The Limited *In Vitro* Lifespan of Human Diploid Cell Strains', *Experimental Cell Research*, Vol. 37 (1965), pp. 614–36.

31   Witkowski, 'Dr. Carrel's Immortal Cells' (1980), pp. 133–5.

32   Leonard Hayflick, interview with the author (20 December 2004). On the use of WI-38 in the Skylab mission, see P.O'B. Montgomery, Jr., J.E. Cook, R.C. Reynolds, J.S. Paul, L. Hayflick, D. Stock, W.W. Schulz, S. Kimsey, R.G. Thirlof, T. Rogers and D. Campbell, 'The Response of Single Human Cells to Zero Gravity', *In Vitro*, Vol. 14 (1978), pp. 165–73.

33   Barbara J. Culliton, 'Grave-Robbing: The Charge Against Four from Boston City Hospital', *Science*, Vol. 186 (1974), pp. 420–3.

34   Culliton, 'Grave-Robbing' (1974).

35   Barbara J. Culliton, 'National Research Act: Restores Training, Bans Fetal Research', *Science*, Vol. 185 (1974), pp. 426–7.

36   Diana S. Hart, 'Fetal Research and Anti-Abortion Politics: Holding Science Hostage', *Family Planning Perspectives*, Vol. 7 (1975), pp. 72–82.

37   Angela Holder and Robert Levine, 'Informed Consent for Research on Specimens Obtained at Autopsy or Surgery: A Case Study in the Overprotection of Human Subjects', *Clinical Research*, Vol. 24 (1976), pp. 68–77; Lewis Corriel,

'The Scientific Responsibilities at Issue', *In Vitro*, Vol. 13, no. 10 (1977), pp. 632–41, on pp. 639–40.

38 Fox and Swazey, *Observing Bioethics* (2008), pp. 128–45.

39 *Browning v Norton Children's Hospital* (1974), 504 SW 2d 713 (Ky CA). See also Bernard M. Dickens, 'The Control of Living Body Materials', *University of Toronto Law Journal*, Vol. 27 (1977), pp. 142–98.

40 *Mokry v University of Texas Health Center in Dallas* (1975) 325 So 2d 479 (Fla Dist Ct of App); Dickens, 'Control of Living Bodily Materials' (1977).

41 Ibid, p. 183.

42 See Stevens, *Bioethics in America* (2003).

43 William J. Winslade, 'An Overview of the Scientist's Responsibilities: Comments by an Attorney', *In Vitro*, Vol. 13, no. 10 (1977), pp. 712–27, on p. 714.

44 Winslade, 'An Overview of the Scientist's Responsibilities' (1977), p. 716.

45 Ibid.

46 Ibid.

47 Ibid, p. 719.

48 B.D. Davis, 'The Social Control of Science', in Alun Milunsky and George Annas (eds), *Genetics and the Law* (New York: Plenum Press, 1975), pp. 301–14.

49 Thomas Hearn, responding to E. Maynard Adams, 'The Ethical Responsibilities at Issue', *In Vitro*, Vol. 13, no. 10 (1977), p. 607.

50 Ronald Nardone, responding to E. Maynard Adams, 'Ethical Responsibilities' (1977), p. 609.

51 William R. Wasserstrom, responding to Marshall Shapo, 'Legal Responsibilities at Issue – Emphasis on Informed Consent', *In Vitro*, Vol. 13, no. 10 (1977), pp. 613–31, on p. 628.

52 Winslade (1977), p. 717.

53 Shapo, responding to Winslade (1977), p. 724.

54 Nicholas Wade, 'Hayflick's Tragedy: The Rise and Fall of a Human Cell Line', *Science*, Vol. 192 (1976), pp. 125–7, on p. 125. See also Stephen Hall, *Merchants of Immorality: Chasing the Dream of Human Life Extension* (Boston: Houghton Mifflin, 2003).

55 Harold M. Schmeck Jr., 'Investigator Says Scientist Sold Cell Specimens Owned by US', *New York Times* (28 March 1976).

56 Wade, 'Hayflick's Tragedy' (1976), p. 127.

57 Leonard Hayflick, 'A Novel Technique for Transforming the Theft of Mortal Human Cells into Praiseworthy Federal Policy', *Experimental Gerontology*, Vol. 33 (1998), pp. 191–207, on p. 196.

58 Constance Holden, 'Hayflick Case Settled', *Science*, Vol. 215 (1982), p. 271.

59 This was extended to cover private firms in 1983. See John Walsh, 'President Tells Agencies to Lower Patent Bars', *Science*, Vol. 219 (1983), pp. 1408–9, on p. 1408.

60 *Diamond v Chakrabarty* 447 US 303, 100 S Ct 2204 US 1980. See also Daniel J. Kevles, '*Diamond v Chakrabarty* and Beyond: The Political Economy of Patenting Life', in Arnold Thackray (ed.), *Private Science: Biotechnology and the Rise of the Molecular Sciences* (Philadelphia: University of Pennsylvania Press, 1998), pp. 65–79.

61 Hayflick, 'A Novel Technique' (1998).

62   Barbara J. Culliton, 'Patient Sues UCLA Over Cell Line', *Science*, Vol. 225 (1984), p. 1458.
63   Office of Technology Assessment, United States Congress, *Ownership of Human Tissues and Cells* (New York: Books for Business, 1987), pp. 45, 50–1.
64   Nuffield Council on Bioethics, *Human Tissue: Ethical and Legal Issues* (London: Nuffield Council on Bioethics, 1995), pp. 85–6.
65   Nicholas Wade, 'University and Drug Firm Battle Over Billion-Dollar Gene', *Science*, Vol. 209 (1984), pp. 1492–4, on p. 1494.
66   H. Phillip Koeffler and David W. Golde, 'Acute Myelogenous Leukemia: A Human Cell Line Responsive to Colony Stimulating Activity', *Science*, Vol. 200 (1978), pp. 1153–4.
67   On the history of interferon, see Toine Pieters, 'Hailing a Wonder Drug: the Interferon', in Willem de Blecourt and Cornelie Usborne (eds), *Cultural Approaches to the History of Medicine* (Basingstoke: Palgrave, 2003), pp. 212–22.
68   Wade, 'Battle Over Billion-Dollar Gene' (1980), p. 1493.
69   Wade (1980), p. 1493.
70   Dorothy Nelkin, *Science as Intellectual Property: Who Controls Scientific Research?* (London: Macmillan Press, 1984), pp. 12–16.
71   Wade (1980), p. 1493.
72   Ibid.
73   The products of these hybridoma cell lines, known as 'monoclonal antibodies', became scientifically important and commercially remunerative tools during the 1980s. For a history of monoclonal antibodies, see Alberto Cambrosio and Peter Keating, *Exquisite Specificity: The Monoclonal Antibody Revolution* (Oxford: Oxford University Press, 1995).
74   Marjorie Sun, 'Scientists Settle Cell Line Dispute', *Science*, Vol. 220 (1983), pp. 393–4.
75   Sun, 'Scientists Settle Cell Line Dispute' (1983), p. 393.
76   Sun (1983), p. 393.
77   Ibid, p. 394.
78   Ibid.
79   Ibid.
80   Ibid.
81   *John Moore v the Regents of the University of California* (1990), 51 Cal 3d 120. For more background on the case, see Waldby and Mitchell, *Tissue Economies* (2007), pp. 88–110; Paul Rabinow, 'Severing the Ties: Fragmentation and Redemption in Late Modernity', in Paul Rabinow, *Essays on the Anthropology of Reason* (Princeton: University of Princeton Press, 1996), pp. 129–53; Hannah Landecker, 'Between Beneficence and Chattel: The Human Biological in Law and Science', *Science in Context*, Vol. 12 (1999), pp. 203–25.
82   Barbara J. Culliton, 'Patient Sues UCLA Over Patent on Cell Line', *Science*, Vol. 225 (1984), p. 1458.
83   Barbara J. Culliton, 'Mo Case Has Its First Court Hearing', *Science*, Vol. 226 (1984), pp. 813–14, on p. 813.
84   Leon E. Rosenberg, 'Using Patient Materials for Product Development: A Dean's Perspective', *Clinical Research*, Vol. 33 (1985), pp. 452–3, on p. 425.
85   Alan E. Otten, 'Researchers' Use of Blood, Bodily Tissues Raises Question About Sharing Profits', *The Wall Street Journal* (29 January 1986).
86   Culliton, 'Mo Case Has Its First Court Hearing' (1984), p. 813.

87 Sandra Blakeslee, 'Patient Sues for Title to Own Cells', *Nature*, Vol. 311 (1984), p. 198.

88 Barbara Culliton, 'Patient Sues UCLA over Patent on Cell Line', *Science*, Vol. 225 (1984), p. 1458.

89 Ivor Royston, 'Cells from Human Patients: Who Owns Them? A Case Report', *Clinical Research*, Vol. 33 (1985), p. 443.

90 Lori B. Andrews, 'My Body, My Property', *Hasting Center Report*, Vol. 16 (1986), pp. 28–38, on p. 37.

91 Andrews, 'My Body, My Property' (1986), p. 29.

92 Andrews (1986), p. 32.

93 Ibid, p. 29.

94 Arthur L. Caplan, 'Blood, Sweat and Tears, and Profits: The Ethics of the Sale and Use of Patient Derived Materials in Biomedicine', *Clinical Research*, Vol. 33 (1985), pp. 448–52.

95 George Annas, 'Outrageous Fortune: Selling Other People's Cells', in George Annas (ed.), *Standard of Care: The Law of American Bioethics* (Oxford: Oxford University Press, 1993), p. 172.

96 William J. Curran, 'Scientific and Commercial Development of Cell Lines: Issues of Property, Ethics and Conflict of Interest', *New England Journal of Medicine*, Vol. 324 (1991), pp. 998–1000.

97 See, for example, Sharon N. Perley, 'From Control Over One's Body to Control Over One's Parts: Extending the Doctrine of Informed Consent', *New York University Law Review*, Vol. 67 (1992), pp. 335–66; Catherine A. Tallerico, 'The Autonomy of the Human Body in the Age of Biotechnology', *University of Colorado Law Review*, Vol. 61 (1990), pp. 659–80.

98 Ian Kennedy, 'What is a Medical Decision? The 1979 Astor Memorial Lecture', reprinted in Ian Kennedy, *Treat Me Right: Essays in Medical Law and Ethics*, (Oxford: Clarendon Press, 1988), pp. 19–31. This argument was made forcefully in Kennedy's 1980 Reith Lectures, reprinted as idem, *The Unmasking of Medicine* (London: Allen and Unwin, 1981).

99 See Brian Salter, *The New Politics of Medicine* (Basingstoke: Palgrave Macmillan, 2004).

100 See Soraya de Chadarevian, *Designs for Life: Molecular Biology After World War II* (Cambridge: Cambridge University Press, 2002), pp. 336–63.

101 Diana Brahams, 'A Disputed Spleen', the *Lancet*, Vol. 332 (1988), pp. 1151–2, on p. 1551.

102 Brahams, 'Disputed Spleen' (1988), p. 1152.

103 Ibid.

104 R. Ian. Freshney, *Culture of Animal Cells: A Manual of Basic Technique* (Chichester: Willey-Liss, 1987), p. 112.

105 Anon., 'Human Tissue as an Alternative in Bio-Medical Research', *Alternatives to Laboratory Animals*, Vol. 14 (1987), pp. 375–80, on p. 375.

106 Anon., 'Human Tissue' (1987), p. 376.

107 J.H. Fentem, 'Conference Report: The Use of Human Tissues in *In Vitro* Toxicology', *Alternatives to Laboratory Animals*, Vol. 21 (1993), pp. 388–9, on p. 389; Diana Brahams 'Ownership of a Spleen', the *Lancet*, Vol. 366 (1990), p. 329.

108 J. Gurney and M. Balls, 'Obtaining Human Tissues for Research and Testing: Practical Problems and Public Attitudes in Britain', in V. Rogiers

(ed.), *Human Cells in In Vitro Pharmaco-Toxicology: Present Status Within Europe* (Brussels: VUB Press, 1993), pp. 315–28.

109   Gurney and Balls, 'Obtaining Human Tissues' (1993), p. 327.

110   Anon., 'Medical and Scientific Uses of Human Tissue', *Alternatives to Laboratory Animals*, Vol. 20 (1992), p. 200.

111   Stephen Lock, 'Toward a National Ethics Committee', *British Medical Journal*, Vol. 300 (1990), pp. 1149–50; see also Sheila Jasanoff, *Designs on Nature: Science and Democracy in Europe and the United States* (Princeton: Princeton University Press, 2005).

112   Nuffield Council on Bioethics, *Human Tissue: Ethical and Legal Issues* (1995), p. iv.

113   Nuffield Council on Bioethics (1995), pp. 9, 24.

114   Ibid, p. 11.

115   Richard Tutton, 'Person, Property and Gift: Exploring the Languages of Tissue Donation', in Richard Tutton and Oonagh Corrigan (eds), *Genetic Databases: Socio-Ethical Issues in the Collection and Use of DNA* (London: Routledge, 2004), pp. 19–39. See also Richard Titmuss, *The Gift Relationship: From Human Blood to Social Policy* (London: Allen and Unwin, 1970).

116   Anon., 'Working Party Speaks Out on the Use of Human Tissue', *British Medical Journal*, Vol. 310 (1995), p. 1159; Chris Broadhead, 'Human Tissue: Ethical and Legal Issues', *Alternatives to Laboratory Animals*, Vol. 23 (1995), p. 435.

117   Nuffield Council on Bioethics (1995), p. 68.

118   Ibid.

119   Ibid.

120   R.D. Start, W. Brown, R.J. Bryant, M.W. Reed, S.S. Cross, G. Kent and J.C.E. Underwood, 'Ownership and Uses of Human Tissue: Does the Nuffield Bioethics Report Accord with Opinion of Surgical Inpatients?', *British Medical Journal*, Vol. 313 (1996), pp. 1366–8.

121   Start et al, 'Ownership and Uses of Human Tissue' (1996), p. 1368.

122   UKCCCR secretariat, 'UKCCCR Guidelines for the Use of Cell Lines in Cancer Research', *British Journal of Cancer*, Vol. 82, no. 9 (2000), pp. 1495–509.

123   Onora O'Neill, 'Medical and Scientific Uses of Human Tissue', *Journal of Medical Ethics*, Vol. 22 (1996), pp. 2–5, on p. 4.

124   UKCCCR, 'Guidelines for the Use of Cell Lines' (2000), p. 1496.

125   Ian R. Stratford, interview with the author (20 May 2005); R. Ian Freshney, interview with the author (8 November 2004).

126   R. Ian Freshney, *Culture of Animal Cells: A Manual of Basic Technique* (Fourth Edition: Chichester: Wiley-Liss, 2000), p. 154.

127   Freshney, *Culture of Animal Cells* (2000), p. 151.

128   Ian R. Freshney, interview with the author (8 November 2004).

129   Salter, *New Politics of Medicine* (2004), p. 57.

130   Salter (2004). See also Clive Seale, Debbie Cavers and Mary Dixon-Woods, 'Commodification of Body Parts: By Medicine or by Media?', *Body and Society*, Vol. 12 (2006), pp. 25–42; Waldby and Mitchell (2007), pp. 37–8.

131   Nigel Bunyan, 'Alder Hey Sold Tissue from Children', the *Daily Telegraph* (27 January 2001).

132 Ken Mason and Graeme Laurie, 'Consent or Property? Dealing with the Body and Its Parts in the Wake of Bristol and Alder Hey', *Modern Law Review*, Vol. 64 (2001), pp. 710–30; Graeme Laurie, *Genetic Privacy: A Challenge to Medico-Legal Norms* (Cambridge: Cambridge University Press, 2002); Jane Wildgoose, 'Who Really Owns Our Bodies?', the *Guardian* (30 January 2001).

133 Wildgoose, 'Who Really Owns Our Bodies?' (2001).

134 Medical Research Council, *Human Tissue and Biological Samples for Use in Research* (London: Medical Research Council, 2001), p. 3.

135 The Wellcome Trust and the Medical Research Council, *Public Perceptions of the Collection of Human Biological Samples* (London: Wellcome Trust and the Medical Research Council, 2000), p. 16.

136 Department of Health, *Human Bodies, Human Choices: The Law on Tissue Retention in England and Wales* (London: HMSO, 2001). The full document is available online through the United Kingdom National Archives, at http://webarchive.nationalarchives.gov.uk/+/www.dh.gov.uk/en/Consultations/Closedconsultations/DH_4081562.

137 See *Human Bodies, Human Choices* (2001), sections 4.4–4.5, 'What the Review Covers', p. 27.

138 Ibid, section 9.7, p. 54.

139 Ibid, section 6.1, p. 34.

140 Ibid, section 17.5–17.6, p. 157.

141 Ibid, section 17.21–17.23, on p. 161.

142 Ibid, section 17.23, p. 157.

143 John R.W. Masters, interview with the author (University College London, 12 January 2004).

144 Ian R. Stratford, interview with the author (University of Manchester, 20 May 2004); Ian R. Freshney, interview with the author (8 November 2004).

145 Madeleine Brindley, 'Research Hit by Organ Scandal', the *Western Mail* (17 December 2002).

146 Brindley, 'Research Hit by Organ Scandal' (2002).

147 *Human Tissue Bill* (Her Majesty's Stationary Office, 2004), p. 8.

148 Colin Blakemore, 'Human Tissue Bill: Views of the Medical Research Council', *MRC Press Release* (26 January 2004).

149 Gaby Hinscliffe and Robin McKie, 'Doctors Beat Curbs on Tissue Research', the *Observer* (6 June 2004).

150 The Human Tissue Authority did not oversee the scientific use of human reproductive materials, such as eggs, sperm and embryos. Research on reproductive material fell under the remit of the Human Fertilization and Embryology Authority, which was established in 1991. At the time of writing, the Conservative and Liberal Democrat coalition government proposes to dismantle the Human Tissue Authority, devolving its work to a new and over-arching Care Quality Commission.

151 L.J.C. Clarke, cited in Muireann Quigley, 'Property: The Future of Human Tissue?', *Medical Law Review*, Vol. 17 (2009), pp. 457–66, on p. 461.

152 L.J.C. Clarke, cited in Nuffield Council on Bioethics consultation paper, *Give and Take? Human Bodies in Medicine and Research* (London: Nuffield Council on Bioethics, April 2010), p. 26.

153 Lawrence (1998), p. 131.
154 See, for example, Andrews and Nelkin, *Body Bazaar* (2001); Kimbrell, *The Human Body Shop* (1997).
155 On how interest groups interpret and represent the 'public', see David Cantor, 'Representing "the Public"', in Steve Sturdy (ed.), *Medicine, Health and the Public Sphere in Britain, 1600–2000* (London: Routledge, 2002), pp. 145–69. A growing literature on the recent emergence of patient rights groups also highlights the complex and often contradictory nature of 'public opinion'. See, for example, Steven Epstein, *Impure Science: AIDS, Activism and the Politics of Knowledge* (Berkeley: University of California Press, 1996); Paul Rabinow, *French DNA: Trouble in Purgatory* (Chicago: University of Chicago Press, 1997).
156 Brindley (2002).
157 Klaus Hoeyer, 'Person, Patent and Property: A Critique of the Commodification Hypothesis', *Biosocieties*, Vol. 2 (2007), pp. 327–48.

Chapter 7   Epilogue: Tissues in Culture

1 Jeremy Laurence, 'Bladder Grown in Lab Hailed as Breakthrough for Organ Transplants', the *Independent* (4 April 2006).
2 Anthony Atala, Stuart B. Bauer, Shay Shoker, James J. Yoo and Alan B. Retik, 'Tissue-Engineered Autologous Bladders for Patients Needing Cytoplasty', the *Lancet*, Vol. 367 (2006), pp. 1241–6.
3 The most famous example was when Charles Vacanti grew an ear on the back of a laboratory mouse at the University of Massachusetts. See Charles A. Vacanti, 'The History of Tissue Engineering', *Journal of Cellular and Molecular Medicine*, Vol. 10 (2006), pp. 569–76.
4 Anon., 'Lab-Grown Bladders "a Milestone"', BBC News Online (3 April 2006). Available online at http://news.bbc.co.uk/1/hi/health/4871540.stm.
5 See Squier, *Liminal Lives* (2004); Waldby and Mitchell, *Tissue Economies* (2007).
6 For more on the promissory discourses surrounding tissue regeneration, see Linda Hogle, 'Life/Time Warranty: Rechargeable Cells and Extendable Lives', in Sarah Franklin and Margaret Lock (eds), *Remaking Life and Death: Toward an Anthropology of the Biosciences* (Santa Fe: School of American Research Press, 2003), pp. 61–97.
7 Oron Catts and Ionat Zurr, 'Growing Semi-Living Sculptures: The Tissue Culture & Art Project', *Leonardo*, Vol. 35 (2002), pp. 365–70, on p. 366. See also Edwina Bartlem, 'Emergence: New Flesh and Life in New Media Art', in Zoe Detsi-Diamanti, Katerina Kitsi-Mitakou and Effie Yiannopoulou (eds), *The Future of Flesh: A Cultural Survey of the Body* (Basingstoke: Palgrave Macmillan, 2009), pp. 155–81.
8 Catts and Zurr, 'Semi-Living Sculptures' (2002), p. 368.
9 Catts and Zurr (2002), p. 366.
10 Ibid.
11 Oron Catts and Ionat Zurr, 'Semi-Living Art, in Eduardo Kac (ed.), *Signs of Life: BioArt and Beyond* (Cambridge, Mass: MIT Press, 2007), pp. 231–49, on p. 232.
12 Artists have regularly used human and animal materials, especially in the second half of the twentieth century, but 'bioart' differs in that its prac-

titioners employ scientific techniques to manipulate discrete biological units – such as DNA, proteins, cells and tissues – and often do so to create transgenic and semi-living entities. See Kac, *Signs of Life* (2007). Of course, this is by no means a recent phenomenon; many techniques and objects have long straddled, and facilitated interplay between, science and art. As Lorraine Daston and Peter Galison demonstrate, anatomical, botanical and zoological illustrations during the eighteenth and nineteenth centuries were drawn by professional artists, to an aesthetic ideal of 'truth to nature'. See Daston and Galison, *Objectivity* (2007), pp. 55–105. And scale models of dinosaurs, human embryos and, more recently, DNA, have simultaneously existed as artistic and scientific objects. See Soraya de Chadarevian and Nick Hopwood (eds), *Models: The Third Dimension of Science* (Stanford: Stanford University Press, 2004).

13 Dr John Hunt, researcher at Liverpool University's UK Centre for Tissue Engineering, quoted in 'Research Intelligence' press release. Available online at http://www.liv.ac.uk/researchintelligence/issue34/skin.htm.

14 See Lorraine Daston (ed.), *Things That Talk: Object Lessons from Art and Science* (New York: Zone Books 2004).

15 Catts and Zurr, 'Semi-Living Art' (2007), p. 231.

16 For more on how experimental objects help constitute networks of material practices, objects, scientific and non-scientific groups, see Bruno Latour, *Reassembling the Social: An Introduction to Actor-Network Theory* (Oxford: Oxford University Press, 2007); idem, *The Pasteurization of France* (Cambridge, Mass: Harvard University Press, 1988).

17 Pickstone, 'Objects and Objectives' (1994), pp. 19–21. Of course, these older objects were not completely displaced by tissue culture and continued to feature in public debates – e.g., in retained organ scandals and recent discussion of 'animal rights'.

18 See Pickstone, *Ways of Knowing* (2000); Turney (1998).

19 See Richardson, 'Fearful Symmetry' (1996); Andrews and Nelkin (2000); idem, (1998); Kimbrell (1997).

20 Bronwyn Parry, *Trading the Genome: Investigating the Commodification of Bio-Information* (New York: Columbia University Press, 2004), pp. 257–9.

21 Onora O'Neill, *Autonomy and Trust in Bioethics* (Cambridge: Cambridge University Press, 2001), p. 128.

22 O'Neill, *Autonomy and Trust in Bioethics* (2001), p. 3.

23 See for example, Sarah Franklin, *Dolly Mixtures: The Remaking of Genealogy* (Durham and London: Duke University Press, 2007); Parry and Gere, 'Contested Bodies' (2006).

24 This argument is expanded, with empirical evidence drawn from patient interviews, in Mary Dixon-Woods, Clare Jackson, Duncan Wilson, Debbie Cavers and Kathy Pritchard-Jones, 'Human Tissue and "the Public": The Case of Childhood Cancer Tumour Banking', *Biosocieties*, Vol. 3 (2008), pp. 57–80. On how the agendas of scientists, patients and social groups are mutually reinforcing, see also Epstein (1996); Susan Lindee, *Moments of Truth in Genetic Medicine* (Baltimore: Johns Hopkins Press, 2005).

25 John V. Pickstone, 'Bureaucracy, Liberalism and the Body in Post-Revolutionary France: Bichat's Physiology and the Paris School of Medicine', *History of*

Science, Vol. 19 (1981), pp. 115–42; Duncan Wilson, 'Historical Keyword: "Tissue"', the *Lancet*, Vol. 373 (2009), p. 109.

26 Michel Serres, *The Five Senses: A Philosophy of Mingled Bodies* (London: Continuum, 2008). See also Steven Connor, *The Book of Skin* (London: Reaktion Books, 2004), p. 29.

27 Rabinow, *French DNA* (1999), p. 180.

28 O'Neill (2001), pp. 158–9.

References

Archives

BBC Written Archives Centre, Caversham Park, Reading
File S, Catalogue number: VR/67/777.

Oxford Brookes University Medical Sciences Video Archive
Professor Sir Henry Harris in Conversation with Sir Gordon Wolstenholme (recorded 6 June 1986): Catalogue number MSVA 012.

University of Cambridge Department of Manuscripts and University Archives
Papers and correspondence of Joseph Needham. Catalogue number: NCUAS 54.3.95/J.234.

University of Liverpool Science Fiction Collection
Amazing Stories (1926–38). Catalogue number: PX1000.A4281.

Wellcome Library for the History and Understanding of Medicine, London, Archives and Manuscripts
Honor Bridget Fell Papers. Catalogue number: CMAC/PP/HBF. File numbers: PP/HBF/B.1; D.21; E.12/3; E.17.
Strangeways Research Laboratory Papers. Catalogue number: CMAC/SA/SRL. File numbers: SA/SRL/A.27; C.3; C.4; E.12; G.1; G.52; H.4; J.2.
Frederick Gordon Spear Papers. Catalogue number: CMAC/PP/FGS. File numbers: CMAC/PP/FGS/C.18.
Alfred Glücksmann Papers. Catalogue number: CMAC/PP/GC. File number: GC/91/C7.

Interviews

L.M.F. Franks (12 May 2004, London).
R. Ian Freshney (8 November 2004, Glasgow).
Leonard Hayflick (23 December 2004, telephone interview).
John M. Masters (12 January 2004, London).
Ian R. Stratford (20 May 2005, Manchester).

Newspapers and Magazines

Daily Express; Daily Mirror; Fort Wayne Gazette; Herald Tribune; Listener; Manchester Guardian; New York Times; Observer; Paris Daily Mail; Radio Times; San Antonio

Light; Scrutiny; Sunday Express; The Times; Sunday Telegraph; Sunday Times; Tit-Bits; Wall Street Journal; Wisconsin State Journal

Legal Cases

Browning v Norton Children's Hospital (1974), 504 SW 2d 713 (Ky CA).
Diamond v Chakrabarty 447 US 303, 100 S Ct 2204 US 1980.
Doodeward v Spence (1908), CLR 406, 414.
John Moore v the Regents of the University of California (1990), 51 Cal 3d 120.
Mokry v University of Texas Health Center at Dallas (1975) 325 So 2d 479 (Fla Dist Ct of App).

Reports and Guidelines

Department of Health, *Human Bodies, Human Choices: The Law on Tissue Retention in England and Wales* (London: Her Majesty's Stationary Office, 2001).
Littlewood, S. (chair), *Report of the Departmental Committee on Experiments on Animals* (London: Her Majesty's Stationary Office, 1965).
Medical Research Council and the Wellcome Trust, *Public Perceptions of the Collection of Human Biological Samples* (London: Medical Research Council, Wellcome Trust, 2000).
Medical Research Council, *Human Tissue and Biological Samples for Use in Research* (London: Medical Research Council, 2001).
Nuffield Council on Bioethics, *Human Tissue: Ethical and Legal Issues* (London: Nuffield Council, 1995).
Nuffield Council on Bioethics, *Give and Take? Human Bodies in Medicine and Research* (London: Nuffield Council, 2010).
Peel, J. (chair), *The Use of Foetuses and Foetal Material for Research: Report of the Advisory Group* (London: Her Majesty's Stationary Office, 1972).
Office of Technology Assessment, United States Congress, *Ownership of Human Tissues and Cells: New Developments in Biotechnology* (New York: Books for Business, 1987).

Books and Articles

Abir-Am, P.G., 'The Assessment of Interdisciplinary Research in the 1930s: The Rockefeller Foundation and Physico-Chemical Morphology', *Minerva*, Vol. 26 (1988), pp. 153–76.
Adams, E.M., 'The Ethical Responsibilities at Issue', *In Vitro*, Vol. 13, no. 10 (1977), pp. 595–612.
Agar, J., 'What Happened in the Sixties?', *British Journal for the History of Science*, Vol. 41 (2008), pp. 567–601.
Anderson, W., *The Collectors of Lost Souls: Turning Kuru Scientists into Whitemen* (Baltimore: Johns Hopkins University Press, 2008).
Andrews, L.B., 'My Body, My Property', *Hasting Center Report*, Vol. 16 (1986), pp. 28–38.
Andrews, L.B. and Nelkin, D., 'Whose Body is it Anyway? Disputes Over Body Tissue in a Biotechnology Age', the *Lancet*, Vol. 351 (1998), pp. 53–7.

Andrews, L.B. and Nelkin, D., *Body Bazaar: The Market for Human Tissue in the Biotechnology Age* (New York: Crown Publishers, 2001).

Annas, G., *Standard of Care: The Law of American Bioethics* (Oxford: Oxford University Press, 1993).

Anon., 'The Research Hospital at Cambridge', the *Lancet*, Vol. 172 (1908), p. 1698.

Anon., 'The Cambridge Hospital for Special Diseases', *British Medical Journal* (1910), p. 522.

Anon., 'Tissue Culture', the *Lancet*, Vol. 201 (1923), p. 858.

Anon., 'X-Ray Effects on Tissue Cultures', the *Lancet*, Vol. 202 (1923), p. 1364.

Anon., 'Recent Developments in Tissue Culture', the *Lancet*, Vol. 204 (1924), pp. 72–3.

Anon., 'Review of *Daedalus, Or Science and the Future*', *Nature*, Vol. 113 (1924), p. 70.

Anon., 'T.S.P. Strangeways', *British Medical Journal* (1927), p. 82.

Anon., 'Obituary: T.S.P. Strangeways', the *Lancet*, Vol. 209 (1927), p. 56.

Anon., *The Strangeways Research Laboratory, Previously the Cambridge Research Hospital* (Cambridge: Heffer and Sons, 1929).

Anon., 'Sir John Bland-Sutton', *Nature*, Vol. 139 (1937), pp. 223–4.

Anon., 'Ariadne', *New Scientist* (26 February 1970), p. 424.

Anon., 'Human Tissue as an Alternative in Biomedical Research', *Alternatives to Laboratory Animals*, Vol. 14 (1987), pp. 375–85.

Anon., 'Medical and Scientific Uses of Human Tissue', *Alternatives to Laboratory Animals*, Vol. 20 (1992), p. 200.

Anon., 'Working Party Speaks Out on the Use of Human Tissue', *British Medical Journal*, Vol. 310 (1995), p. 1159.

Appadurai, A., *The Social Life of Things: Commodities in Cultural Perspective* (Cambridge: Cambridge University Press, 1986).

Appel, T., *Shaping Biology: The National Science Foundation and American Biological Research*, 1945–1975 (Baltimore: Johns Hopkins University Press, 2000).

Armstrong, T., *Modernism, Technology and the Body: A Cultural Study* (Cambridge: Cambridge University Press, 1998).

Armstrong, T., *Modernism* (Cambridge: Polity Press, 2005).

Ashcroft, R. and Dixon-Woods, M., 'Regulation and the Social Licence for Medical Research', *Medical Health Care and Philosophy*, Vol. 11 (2008), pp. 381–91.

Ashley, M., *The Time Machines: The Story of Science-Fiction Pulp Magazines from the Beginning to 1950* (Liverpool: Liverpool University Press, 2000).

Atala, A.T., Bauer, S.B., Shoker, S., Yoo, J.J. and Retik, A.B., 'Tissue-Engineered Autologous Bladders for Patients Needing Cytoplasty', the *Lancet*, Vol. 367 (2006), pp. 1241–6.

Austoker, J., 'Walter Morley Fletcher and the Origins of a Basic Biomedical Research Policy', in Austoker J. and Bryder, L. (eds), *Historical Perspectives on the Role of the MRC* (Oxford: Oxford University Press, 1989), pp. 23–35.

Bang, F.B., 'History of Tissue Culture at Johns Hopkins', *Bulletin of the History of Medicine*, Vol. 51 (1977), pp. 516–37.

Bartlem, E., 'Emergence: New Flesh and Life in New Media Art', in Detsi-Diamanti, Z., Kitsi-Mitakou, K. and Yiannopoulou, E. (eds), *The Future of Flesh: A Cultural Survey of the Body* (London: Palgrave Macmillan, 2009), pp. 155–81.

Bernal, *The World, The Flesh and the Devil: An Inquiry into the Future of the Three Enemies of the Rational Soul* (Second Edition: London: Jonathan Cape, 1970).

Birkenhead, F.E., *The World in 2030 A.D.* (London: Hodder and Stoughton, 1930).

Blakemore, C., 'Human Tissue Bill: Views of the Medical Research Council' (Press Release: 26 January 2004).

Blakeslee, S., 'Patient Sues for Title to Own Cells', *Nature*, Vol. 311 (1984), p. 198.

Blom, P., *The Vertigo Years: Change and Culture in the West, 1900–1914* (London: Weidenfield and Nicolson, 2008).

Bookchin, D. and Schumacher, J., *The Virus and the Vaccine: The True Story of a Cancer-Causing Virus and the Millions of Americans Exposed* (New York: St. Martin's Press, 2004).

Boon, T., *Films of Fact: A History of Science in Documentary Films and Television* (London and New York: Wallflower Press, 2008).

Bowler, P., *Science for All: The Popularization of Science in Early Twentieth-Century Britain* (Chicago: University of Chicago Press, 2009).

Brachet, A., 'Recherches sur le determinisme heredetaire de l' oeuf des Mammiferes. Development *in vitro* de jeunes vesicules blastodermiques de Lapin', *Archives de Biologie (Liege)*, Vol. 28 (1913), pp. 447–503.

Brahams, D., 'Medicine and the Law: A Disputed Spleen', the *Lancet*, Vol. 332 (1988), pp. 1151–2.

Brahams, D., 'Ownership of a Spleen', the *Lancet*, Vol. 336 (1990), p. 239.

Brookes, B., *Abortion in England, 1900–1967* (London: Croom Helm, 1988).

Brown, R.W. and Henderson, J.H.M., 'The Mass Production and Distribution of HeLa Cells at Tuskegee Institute, 1953–1955', *Journal of the History of Medicine*, Vol. 38 (1983), pp. 695–715.

Bryder, L., 'Tuberculosis and the MRC', in Austoker J. and Bryder L. (eds), *Historical Perspectives on the Role of the MRC* (Oxford: Oxford University Press, 1989), pp. 3–23.

Bud, R., 'Penicillin and the New Elizabethans', *British Journal for the History of Science*, Vol. 31 (1998), pp. 305–33.

Bud, R., *Penicillin: Triumph and Tragedy* (Oxford: Oxford University Press, 2007).

Burhop, E.H.S., 'The British Society for Social Responsibility in Science', *Physics Education*, Vol. 6 (1971), pp. 140–2.

Burrows, M., 'The Cultivation of Human Cancer Cells In Vitro', *Medical Record*, Vol. 86 (1914), p. 649.

Burrows, M., Burns, J.E. and Suzuki, Y., 'Studies on the Growth of Cells: The Cultivation of Bladder and Prostatic Tumours Outside of the Body', *Journal of Urology*, Vol. 1 (1917), pp. 3–15.

Busby, H., 'Informed Consent: The Contradictory Ethical Safeguards in Pharmacogenetics', in Tutton, R. and Corrigan O. (eds), *Genetic Databases: Socio-Legal Issues in the Collection and Use of DNA* (London: Routledge, 2004), pp. 78–97.

Bynum, C.W., *Metamorphosis and Identity* (New York: Zone Books, 2005).

Cambrosio, A. and Keating, P., *Exquisite Specificity: The Monoclonal Antibody Revolution* (Oxford: Oxford University Press, 1995).

Cantor, D., 'The Definition of Radiobiology' (University of Lancaster PhD thesis, 1987).

Cantor, D., 'The MRC's Support for Experimental Radiology During the Inter-War Years', in Austoker, J. and Bryder, L., *Historical Perspectives on the Role of the MRC* (Oxford: Oxford University Press, 1989), pp. 181–204.

Cantor, D., 'Representing "the Public": Medicine, Charity and Emotion in Twentieth-Century Britain', in Sturdy, S. (ed.), *Medicine, Health and the Public Sphere in Britain, 1600–2000* (London: Routledge, 2000), pp. 145–69.

Caplan, A.L., 'Blood, Sweat and Tears, and Profits: The Ethics of the Sale and Use of Patient Derived Materials in Biomedicine', *Clinical Research*, Vol. 33 (1985), pp. 448–52.

Carleton, H.M., 'Tissue Culture: A Critical Summary', *British Journal of Experimental Biology*, Vol. 1 (1923), pp. 131–51.

Carrel, A. and Burrows, M., 'Cultivation of Adult Tissues and Organs Outside of the Body', *Journal of the American Medical Association*, Vol. 55 (1910), pp. 1379–81.

Carrel, A. and Burrows, M., 'Human Sarcoma Cultivated Outside of the Body', *Journal of the American Medical Association*, Vol. 55 (1910), p. 1732.

Carrel, A. and Burrows, M., 'Cultivation of Tissue In Vitro and Its Technique', *Journal of Experimental Medicine*, Vol. 13 (1911), pp. 387–96.

Carrel, A. and Burrows, M., 'Cultivation In Vitro of Malignant Tumours', *Journal of Experimental Medicine*, Vol. 13 (1911), pp. 571–5.

Carrel, A., 'On the Permanent Life of Tissues Outside of the Organism', *Journal of Experimental Medicine*, Vol. 15 (1912), pp. 276–9.

Catts, O. and Zurr, I., 'Growing Semi-Living Sculptures: The Tissue Culture and Art Project', *Leonardo*, Vol. 35 (2002), pp. 355–70.

Catts, O. and Zurr, I., 'Semi-Living Art', in Kac, E. (ed.), *Signs of Life: BioArt and Beyond* (Cambridge, Mass: MIT Press, 2007), pp. 231–49.

Clark, R.W., *J.B.S.: The Life and Work of J.B.S. Haldane* (Oxford: Oxford University Press, 1984).

Clarke, A.E., 'Research Materials and Reproductive Science in the United States, 1910–40', in Geison, G.L. (ed.), *Physiology in the American Context, 1850–1940* (Baltimore: Williams and Wilkins, 1987), pp. 323–50.

Connor, S., *The Book of Skin* (London: Reaktion Books, 2004).

Conyers, B., 'Animal Experiments', *New Scientist*, Vol. 63 (1974), p. 757.

Cooter, R., 'The Ethical Body', in Pickstone, J. and Cooter, R. (eds), *Medicine in the Twentieth Century* (Amsterdam: Harwood Academic Press, 2000), pp. 451–67.

Corner, G.W., *The Seven Ages of a Medical Scientist: An Autobiography* (Pennsylvania: University of Pennsylvania Press, 1981).

Corriel, L.L., 'The Scientific Responsibilities at Issue', *In Vitro*, Vol. 13, no. 10 (1977), pp. 632–41.

Corrigan, O., 'Informed Consent: The Contradictory Ethical Safeguards in Pharmacogenetics', in Tutton, R. and Corrigan, O. (eds), *Genetic Databases: Socio-ethical Issues in the Collection and Use of DNA* (London: Routledge, 2004), pp. 78–97.

Corner, G.W., *A History of the Rockefeller Institute, 1901–1953: Origins and Growth* (New York: Rockefeller Institute Press, 1965).

Crossley, R., 'Olaf Stapledon and the Idea of Science Fiction', *Modern Fiction Studies*, Vol. 32 (1986), pp. 21–42.

Crossley, R., *Olaf Stapledon: Speaking for the Future* (Liverpool: University of Liverpool Press, 1994).

Culliton, B.J., 'Grave Robbing: The Charge Against Four from Boston City Hospital', *Science*, Vol. 186 (1974), pp. 420–3.

Culliton, B.J., 'National Research Act: Restores Training, Bans Fetal Research', *Science*, Vol. 185 (1974), pp. 426–7.

Culliton, B.J., 'Fetal Research: The Case History of a Massachusetts Law', *Science*, Vol. 187 (1975), pp. 240–1.

Culliton, B.J., 'Patient Sues UCLA over Patent on Cell Line', *Science*, Vol. 225 (1984), p. 1458.

Culliton, B.J., 'Mo Case Has First Court Hearing', *Science*, Vol. 226 (1984), pp. 813–14.

Curran, W.J., 'Scientific and Commercial Development of Cell Lines: Issues of Property, Ethics, and Conflict of Interest', *New England Journal of Medicine*, Vol. 324, no. 14 (1991), pp. 998–1000.

Daston, L. and Park, K., *Wonders and the Order of Nature, 1150–1750* (New York: Zone Books, 1998).

Daston, L. (ed.), *Things That Talk: Object Lessons from Art and Science* (New York: Zone Books, 2004).

Daston, L. and Gailson, P., *Objectivity* (New York: Zone Books, 2007).

Dean, M., *Governmentality: Power and Rule in Modern Society* (London: Sage Publications, 2010).

De Chadarevian, S., *Designs for Life: Molecular Biology After World War II* (Cambridge: Cambridge University Press, 2002).

De Chadarevian, S. and Hopwood, N. (eds), *Models: The Third Dimension of Science* (Stanford: Stanford University Press, 2004).

Dickens, B.M., 'The Control of Living Body Materials', *University of Toronto Law Journal*, Vol. 27 (1977), pp. 142–98.

Dingle, J.T. and Fell, H.B.F., 'Famous Laboratories: Strangeways Research Laboratory', *Biologist*, Vol. 31 (1984), pp. 191–7.

Dixon-Woods, M., Jackson, C., Wilson, D., Cavers, D. and Pritchard-Jones, K., 'Human Tissue and "the Public": The Case of Childhood Cancer Banking', *Biosocieties*, Vol. 3 (2008), pp. 57–81.

Douglas, M., *Purity and Danger* (London: Routledge, 2004).

Drew, A.H., 'Three Lectures on the Cultivation of Tissues and Tumours In Vitro', the *Lancet*, Vol. 201 (1923), pp. 834–5.

Dronamrjau, K.R. (ed.), *Haldane's Daedalus Revisited* (Oxford: Oxford University Press, 1995).

Dyer, O., 'Working Party Speaks Out on the Use of Human Tissue', *British Medical Journal*, Vol. 310 (1995), p. 1159.

Edwards, R.G., Bavister, B.V. and Steptoe, P.C., 'Early Stages of Fertilization *in vitro* of Human Oocytes Matured *in vitro*', *Nature*, Vol. 221 (1969), pp. 632–5.

Edwards, R.G., 'Aspects of Human Reproduction', in Fuller, W. (ed.), *The Social Impact of Modern Biology* (London: Routledge and Kegan Paul, 1971), pp. 108–22.

Enders, J.F., Weller, T.H. and Robbins, F.C., 'Cultivation of the Lansing Strain of Poliomyelitis Virus in Cultures of Various Human Embryonic Tissues', *Science*, Vol. 109 (1949), pp. 85–7.

Enders, J.F. and Peebles, T.C., 'Propagation in Tissue Culture of Cytopathogenic Agents from Patients with Measles', *Proceedings of the Society for Experimental Biology and Medicine*, Vol. 86 (1954), pp. 277–86.

Epstein, S., *Impure Science: AIDS, Activism, and the Politics of Knowledge* (Berkley: University of California Press, 1996).

Evans, V.J., Hawkins, N.M., Westfall, B.B. and Earle, W.R., 'Studies on Culture Lines Derived from Mouse Liver Parenchymatous Cells Grown in Long Term Tissue Culture', *Cancer Research*, Vol. 18 (1958), pp. 261–6.

Fell, H.B.F., 'The Development *In Vitro* of the Isolated Otocyst of the Embryonic Fowl', *Archiv für Experimentelle Zellforschung*, Vol. 7 (1928), pp. 69–81.

Fell, H.B.F., 'Tissue Culture: The Advantages and Limitations as a Research Method', *British Journal of Radiology*, Vol. 8 (1935), pp. 27–31.

Fell, H.B.F., 'Cell Biology', in Fell, Spear, F.G. and Strangeways, E.D., *The History of the Strangeways Research Laboratory, 1912–1962* (Cambridge, 1962), pp. 6–12.

Fell, H.B.F., 'The Linacre Lecture, 1969: Cells in Captivity: Past, Present and Future', *Journal of the Medical Women's Federation*, Vol. 52 (1970), pp. 32–48.

Fell, H.B.F., 'Tissue Culture and Its Contribution to Biology and Medicine', *Journal of Experimental Biology*, Vol. 57 (1972), pp. 1–13.

Fell, H.B.F., 'The Development of Organ Culture', in Balls, M. and Monnickendam, M. (eds), *British Society for Cell Biology Symposium 1: Organ Culture in Biomedical Research: Festschrift for Dame Honor Fell, FRS* (Cambridge: Cambridge University Press, 1976), pp. 1–13.

Feller, A.E., Enders, J.F. and Weller, T.H., 'The Prolonged Coexistence of Vaccinia Virus in High Titre and Living Cells in Roller Tube Cultures of Chick Embryo Tissues', *Journal of Experimental Medicine*, Vol. 72 (1940), pp. 367–88.

Fentern, J.H., 'Conference Reports: "The Use of Human Tissues *In Vitro* Toxicology"', *Alternatives to Laboratory* Animals, Vol. 21 (1993), pp. 388–9.

Fjelde, A., 'Human Tumor Cells in Tissue Culture', *Cancer*, Vol. 8 (1955), pp. 845–51.

Flagg, F., 'The Machine Man of Ardathia', in Greenberg, M.H. (ed.), *Amazing Science Fiction Anthology: The Wonder Years, 1926–1935* (London: TSR UK, 1987), pp. 77–95.

Fox, R., 'The Evolution of American Bioethics: A Sociological Perspective', in Weisz, G. (ed.), *Social Science Perspectives on Medical Ethics* (Dordrecht: Kluwer Academic Publishers, 1990), pp. 201–17.

Fox, R. and Swazey, J., *Observing Bioethics* (Oxford: University of Oxford Press, 2008).

Franklin, S., 'Ethical Biocapital: New Strategies of Cell Culture', in Franklin, S. and Lock, M. (eds), *Remaking Life and Death: Toward an Anthropology of the Biosciences* (Santa Fe: School of American Research Press, 2003), pp. 97–129.

Franklin, S., *Dolly Mixtures: The Remaking of Genealogy* (Durham and London: Duke University Press, 2007).

Franks, L.M.F., 'Summary and Future Developments', in Balls, M. and Monnickendam, M. (eds), *Organ Culture in Biomedical Research: Festschrift for Dame Honor Fell* (Cambridge: Cambridge University Press, 1976), pp. 546–57.

French, R.D., *Antivivisection and Medical Science in Victorian Society* (Princeton: Princeton University Press, 1975).

Freshney, R.I., *Culture of Animal Cells: A Manual of Basic Technique* (Second Edition: Chichester: Wiley-Liss Inc., 1987).

Freshney, R.I., *Culture of Animal Cells: A Manual of Basic Technique* (Fourth Edition: Chichester: Wiley-Liss Inc., 2000).

Friedman, M. and Friedland, G.W., *Medicine's 10 Greatest Discoveries* (London: Yale University Press, 1998).

Fujimara, J., *Crafting Science: A Sociohistory of the Quest for the Genetics of Cancer* (Cambridge: Harvard University Press, 1997).

Fukuyama, F., *Our Posthuman Future: Consequences of the Biotechnology Revolution* (London: Profile Books, 2002).

Fund for Replacements of Animals in Medical Experiments, *Is the Experimental Animal Obsolete?* (London: FRAME, 1970).

Garrow, D.J., *Liberty and Sexuality: The Right to Privacy and the Making of Roe v Wade* (Oxford: Maxwell Macmillan International, 1994).

Geison, G.L., *Michael Foster and the Cambridge School of Physiology: The Scientific Enterprise in Late Victorian Society* (Princeton: Princeton University Press, 1978).

Gelhorn, A., 'Medical Ethics – So What's the Story?', *In Vitro*, Vol. 13, no. 10 (1977), pp. 588–95.

Glücksmann, A., 'Studies on Bone Mechanics In Vitro: II. The Role of Tension and Pressure in Chondrogenesis', *Anatomical Record*, Vol. 73 (1939), pp. 39–50.

Glücksmann, A., 'Preliminary Observations on the Quantitative Examination of Human Biopsy Material Taken from an Irradiated Carcinoma', *British Journal of Radiology*, Vol. 14 (1941), pp. 187–98.

Glücksmann, A., 'Quantitative Histological Analysis of Radiation-Effects in Human Carcinomata', *British Medical Bulletin*, Vol. 4 (1946), pp. 26–30.

Glücksmann, A., 'Cell Deaths in Normal Vertebrate Ontology', *Biological Reviews*, Vol. 26 (1951), pp. 56–89.

Glücksmann, A., 'Mitosis and Degeneration in the Morphogenesis of the Human Foetal Lung *In Vitro*', *Zeitscrhift und Mikroskopiche Anatomie*, Vol. 64 (1964), pp. 101–10.

Gold, M., *A Conspiracy of Cells: One Woman's Immortal Legacy and the Medical Scandal It Caused* (Albany: State University of New York Press, 1986).

Gregory, J. and Miller, S., *Science in Public: Communication, Culture and Credibility* (Cambridge, Mass: Perseus Publishing, 1998).

Gurney, J. and Balls, M., 'Obtaining Human Tissues for Research and Testing: Practical Problems and Public Attitudes in Britain', in Rogiers, V., Sonck, W., Shepard, E. and Vercruysse, A. (eds), *Human Cells in In Vitro Pharmaco-Toxicology: Present Status Within Europe* (Brussels: VUB Press, 1993), pp. 315–28.

Haldane, J.B.S., *Daedalus, Or Science and the Future* (London: Kegan Paul, Trench, Trubner & Co., Ltd., 1924).

Hall, L.A., 'Chloe, Olivia, Isabel, Letitia, Harriette, Honor, and Many More: Women in Medicine and Biomedical Science, 1914–1945', in Oldfield, S. (ed.), *This Working-Day World: Women's Lives and Culture(s) in Britain, 1914–1945* (London: Taylor and Francis, 1994), pp. 192–202.

Hall, L.A., 'Illustrations from the Wellcome Institute Library: The Strangeways Research Laboratory: Archives in the Contemporary Medical Archives Centre', *Medical History*, Vol. 40 (1996), pp. 231–8.

Hall, S.S., *Merchants of Immortality: Chasing the Dream of Human Life Extension* (Boston: Houghton Mifflin Co., 2003).

Haraway, D.J., *Crystals, Fabrics and Fields: Metaphors of Organicism in Twentieth-Century Biology* (New Haven: Yale University Press, 1976).

Harris, H., 'Hybrid Cells from Mouse and Man', *New Scientist* (18 February 1965), pp. 420–2.

Harris, H., *Cell Fusion: The Dunham Lectures* (Oxford: Clarendon Press, 1970).

Harris, H., *The Balance of Improbabilities: A Scientific Life* (Oxford: Oxford University Press, 1987).

Harris, H., *The Cells of the Body: A History of Somatic Cell Genetics* (Cold Spring Harbor: Cold Spring Harbor Press, 1995).

Harris, H. and Watkins, J.F., 'Hybrid Cells Derived from Mouse and Man: Artificial Heterokaryons of Mammalian Cells from Different Species', *Nature*, Vol. 205 (1965), pp. 640–6.

Harris, J., *On Cloning* (London: Routledge, 2004).

Harris, J.E., 'Structure and Function in the Living Cell', in Johnson, M.L. and Abercrombie, M. (eds), *New Biology, Five* (London: Penguin, 1948), pp. 26–47.

Harrison, R.G., 'Experimental Biology and Medicine', *Physician and Surgeon*, Vol. 34 (1912), pp. 49–65.

Hart, D.S., 'Fetal Research and Antiabortion Politics: Holding Science Hostage', *Family Planning Perspectives*, Vol. 7 (1975), pp. 72–82.

Harvey, A.M., 'Johns Hopkins – The Birthplace of Tissue Culture: The Story of Ross G. Harrison, Warren H. Lewis and George O. Gey', *The Johns Hopkins Medical Journal*, Vol. 136 (1975), pp. 142–9.

Hayflick, L. and Moorhead, P., 'The Serial Cultivation of Human Diploid Cell Strains', *Experimental Cell Research*, Vol. 25 (1961), pp. 585–21.

Hayflick, L., Plotkin, S., Norton, T.W. and Koprowski, H., 'Preparation of Poliovirus in a Human Fetal Diploid Strain', *American Journal of Hygiene*, Vol. 75 (1962), pp. 240–58.

Hayflick, L., 'The Limited *In Vitro* Lifespan of Human Diploid Cell Strains', *Experimental Cell Research*, Vol. 37 (1965), pp. 614–36.

Hayflick, L., 'The Coming of Age of WI-38', in Maramorosch, K. (ed.), *Advances in Cell Culture: Volume 3* (Orlando: Academic Press, 1984), pp. 303–16.

Hayflick, L., 'A Novel Technique for Transforming the Theft of Mortal Human Cells into Praiseworthy Federal Policy', *Experimental Gerontology*, Vol. 33 (1998), pp. 191–207.

Hedgecoe, A., 'A Form of Practical Machinery: The Origins of Research Ethics Committees in the UK, 1967–1972', *Medical History*, Vol. 53 (2009), pp. 331–50.

Hoeyer, K., 'Person, Patent and Property: A Critique of the Commodification Hypothesis', *Biosocieties*, Vol. 2 (2007), pp. 327–48.

Hogle, L.F., 'Life/Time Warranty Rechargeable Cells and Extendable Lives', in Franklin S. and Lock, M. (eds), *Remaking Life and Death: Toward an Anthropology of the Biosciences* (School of American Research Press, 2003), pp. 61–97.

Holden, C., 'Hayflick Case Settled', *Science*, Vol. 215 (1982), p. 271.

Holder, A.R. and Levine, R.J., 'Informed Consent for Research on Specimens Obtained at Autopsy or Surgery: A Case Study in the Overprotection of Human Subjects', *Clinical Research*, Vol. 24 (1976), pp. 68–77.

Hopwood, N., 'Producing Development: The Anatomy of Human Embryos and the Norms of Wilhelm His', *Bulletin of the History of Medicine*, Vol. 74, no. 1 (2000), pp. 29–79.

Hopwood, N., 'Embryology', in Bowler, P.J. and Pickstone, J.V.P. (eds), *The Cambridge History of Science, Volume 6: The Modern Biological and Earth Sciences* (Cambridge: Cambridge University Press, 2009), pp. 287–316.

Huxley, A., 'To the Puritan All Things are Impure', in Huxley, A., *Music at Night & Other Essays* (London: Chatto and Windus, 1931), pp. 173–84.

Huxley, A., 'Economists, Scientists and Humanists', in Adams, M. (ed.), *Science in the Changing World* (London: Allen and Unwin, 1932), pp. 208–23.

Huxley, A., *Brave New World* (London: Flamingo Modern Classics, 1994).

Huxley, A., *Antic Hay* (London: Vintage, 2004).

Huxley, J., 'Searching for the Elixir of Life', *Century Illustrated Monthly*, Vol. 103 (1922), pp. 621–9.

Huxley, J., 'Progress, Biological and Other', in Huxley, J., *Essays of a Biologist* (London: Pelican Books, 1923), pp. 17–65.

Huxley, J., 'The Life Cycle', in Huxley, J., *Essays in Popular Science* (London: Chatto and Windus, 1926), pp. 75–105.

Huxley, J., 'Elixir Vitae' in Huxley, J., *Essays in Popular Science* (London: Chatto and Windus, 1926), pp. 128–30.

Huxley, J., 'Man as a Relative Being', in Adams, M. (ed.), *Science in the Changing World* (London: Allen and Unwin, 1932), pp. 110–30.

Huxley, J., 'The Tissue Culture King', in Cronklin, G. (ed.), *Great Science Fiction by Scientists* (New York: Collins Books, 1970), pp. 348–65.

Jackson, K., *George Newnes and the New Journalism in Britain, 1880–1910* (Aldershot: Ashgate Publishing, 2001).

Jacobs, J.P., Jones, C.M. and Baille, J.P., 'The Characteristics of a Human Diploid Cell Designated MRC-5', *Nature*, Vol. 227 (1970), pp. 168–70.

Jasanoff, S. (ed.), *States of Knowledge: The Co-Production of Science and the Social Order* (London and New York: Routledge, 2004).

Jasanoff, S., *Designs on Nature: Science and Democracy in Europe and the United States* (Princeton: Princeton University Press, 2005).

Jonsen, A., *The Birth of Bioethics* (Oxford: Oxford University Press, 1998).

Jordonova, L., *Sexual Visions: Images of Gender in Science and Medicine Between the Eighteenth and Twentieth Centuries* (Madison: University of Wisconsin Press, 1989).

Keller, D.H., 'A Biological Experiment', in Keller, D.H., *Tales from Underwood* (Jersey: Spearman Press, 1952), pp. 135–52.

Kelty, C. and Landecker, H., 'A Theory of Animation: Cells, L-systems and Film', *Grey Room*, Vol. 17 (2004), pp. 30–63.

Kennedy, I., *The Unmasking of Medicine* (London: Allen and Unwin, 1981).

Kennedy, I., 'What is a Medical Decision? The 1979 Astor Memorial Lecture', reprinted in Kennedy, I., *Treat Me Right: Essays in Medical Law and Ethics* (Oxford: Clarendon Press, 1988), pp. 19–31.

Kern, S., *The Culture of Time and Space, 1880–1918* (Cambridge, Mass: Harvard University Press, 2003).

Kevles, D.J., 'Diamond v. Chakrabarty and Beyond: The Political Economy of Patenting Life', in Thackray, A. (ed.), *Private Science: Biotechnology and the Rise of the Molecular Sciences* (Philadelphia: University of Pennsylvania Press, 1998), pp. 65–79.

Kevles, D.J., *In the Name of Eugenics: Genetics and the Uses of Human Heredity* (Cambridge, Mass: Harvard University Press, 2004).

Kimbrell, A., *The Human Body Shop: The Cloning, Engineering and Marketing of Life* (Washington D.C.: Regenery Press, 1997).

Kirk, R.G.W., *Reliable Animals, Responsible Scientists: Constructing Standard Laboratory Animals in Britain c.1919–1972* (University of London PhD thesis, 2006).

Klotzko, J.A., *A Clone of Your Own? The Art and Science of Cloning* (Oxford: Oxford University Press, 2004).

Koeffler, H.P. and Golde, D.W., 'Acute Myelogenous Leukemia: A Human Cell Line Responsive to Colony-Stimulating Activity', *Science*, Vol. 200 (1978), pp. 1153–4.

Koprowski, H., 'Live Poliomyelitis Virus Vaccines: Present Status and Problems for the Future', *Journal of the American Medical Association*, Vol. 178 (1961), pp. 1151–5.

Lambert, R.A., 'Technique of Cultivating Human Tissues In Vitro', *Journal of Experimental Medicine*, Vol. 24 (1916), pp. 367–72.

Lambert, R.A., 'The Comparative Resistance of Bacteria and Human Tissue Cells to Certain Common Antiseptics', *Journal of Experimental Medicine*, Vol. 24 (1916), pp. 683–8.

Landecker, H., 'Between Beneficence and Chattel: The Human Biological in Law and Science', *Science in Context*, Vol. 12 (1999), pp. 203–25.

Landecker, H., 'Immortality, In Vitro: A History of the HeLa Cell Line', in Brodwin, P. (ed.), *Biotechnology and Culture: Bodies, Anxieties, Ethics* (Bloomington: Indiana University Press, 2000), pp. 53–72.

Landecker, H., 'New Times for Biology: Ross Harrison and the Development of Cellular Life *In Vitro*', *Studies in the History and Philosophy of Biological and Biomedical Sciences*, Vol. 33 (2002), pp. 667–94.

Landecker, H., 'On Beginning and Ending with Apoptosis: Cell Death and Biomedicine', in Franklin, S. and Lock, M. (eds), *Remaking Life and Death: Toward an Anthropology of the Biosciences* (Santa Fe: School of American Research Press, 2003), pp. 23–61.

Landecker, H., 'Building "A New Type of Body in which to Grow a Cell": Tissue Culture at the Rockefeller Institute, 1910–1914', in Stapledon, D. (ed.), *Creating a Tradition of Biomedical Research: Contributions to the History of the Rockefeller University* (New York: Rockefeller University Press, 2004), pp. 151–74.

Landecker, H., 'Cellular Features: Microcinematography and Film Theory', *Critical Inquiry*, Vol. 31 (2005), pp. 903–37.

Landecker, H., 'Microcinematography and the History of Science and Film', *Isis*, Vol. 97 (2007), pp. 121–32.

Landecker, H., *Culturing Life: How Cells Became Technologies* (Cambridge, Mass Harvard University Press, 2007).

Lansbury, C., *The Old Brown Dog: Women, Workers, and Vivisection in Edwardian England* (Madison: University of Wisconsin Press, 1985).

Larsh, H.W., Silberg, S.L. and Hinton, A., 'Use of the Tissue Culture Method in Evaluating Fungal Agents', *Antibiotics Annual*, Vol. (1957), pp. 918–22.

Lasnitzki, I., 'The Effect of 3-4 Benzpyrene on Human Foetal Lung Grown *In Vitro*', *British Journal of Cancer*, Vol. 10 (1956), pp. 510–16.

Lasnitzki, I., 'Observations of the Effects of Condensates from Cigarette Smoke on Foetal Lung *In Vitro*', *British Journal of Cancer*, Vol. 12 (1958), pp. 547–52.

Lasnitzki, I., 'The Effect of a Hydrocarbon-enriched Fraction of Cigarette Smoke Condensate on Human Foetal Lung Grown *In Vitro*', *Cancer Research*, Vol. 28 (1968), pp. 510–16.

Latour, B., *The Pasteurization of France* (Cambridge, Mass: Harvard University Press, 1988).

Latour, B., *Reassembling the Social: An Introduction to Actor-Network Theory* (Oxford: Oxford University Press, 2007).

Laurie, G., *Genetic Privacy: A Challenge to Medico-Legal Norms* (Cambridge: Cambridge University Press, 2002).

Lawrence, S.C., 'Beyond the Grave – the Use and Meaning of Human Body Parts: A Historical Introduction', in Weir, R.F. (ed.), *Stored Tissue Samples: Ethical, Legal, and Public Policy Implications* (Iowa: University of Iowa Press, 1998), pp. 111–43.

Leach, E., *A Runaway World? The Reith Lectures 1967* (London: British Broadcasting Corporation, 1967).

Lederer, S.E., *Subjected to Science: Human Experimentation in America Before the War* (Baltimore: Johns Hopkins University Press, 1995).

Lederer, S.E., *Flesh and Blood: Organ Transplantation and Blood Transfusion in Twentieth-Century America* (Oxford: Oxford University Press, 2008).

Lederer, S.E., 'Experimentation and Ethics', in Bowler, J. and Pickstone, J.V. (eds), *The Cambridge History of Science, Volume Six: The Modern Biological and Earth Sciences* (Cambridge: Cambridge University Press, 2009), pp. 583–600.

LeMahieu, D.L., *A Culture for Democracy: Mass Communication and the Cultivated Mind in Britain Between the Wars* (Oxford: Clarendon Press, 1988).

Lindee, S., *Moments of Truth in Genetic Medicine* (Baltimore: Johns Hopkins University Press, 2005).

Lock, M., *Twice Dead: Organ Transplants and the Reinvention of Death* (Berkley: University of California Press, 2002).

Lock, S. 'Toward a National Ethics Committee', *British Medical Journal*, Vol. 300 (1990), pp. 1149–50.

Locke, J., *Two Treatises of Government* (Cambridge: Cambridge University Press, 1986).

Losee, J.R. and Ebeling, A.H., 'The Cultivation of Human Tissue In Vitro', *Journal of Experimental Medicine*, Vol. 19 (1914), pp. 593–602.

Losee, J.R. and Ebeling, A.H., 'The Cultivation of Human Sarcomatous Tissue In Vitro', *Journal of Experimental Medicine*, Vol. 20 (1914), pp. 140–8.

Ludovici, A., *Lysistrata: Woman's Future and Future Woman* (London: Kegan Paul, Trench, Trubner & Co., 1924).

Lynch, M., 'Sacrifice and the Transformation of the Animal Body into a Scientific Object: Laboratory Culture and Ritual Practice in the Neurosciences', *Social Studies of Science*, Vol. 18 (1988), pp. 265–89.

Maienschein, J., *Transforming Traditions in American Biology, 1880–1915* (Baltimore: Johns Hopkins University Press, 1991).

Maienschein, J., *Whose View of Life? Embryos, Cloning and Stem Cells* (London: Harvard University Press, 2003).

Mason, J.K. and Laurie, G.T., 'Consent or Property? Dealing with the Body and its Parts in the Shadow of Bristol and Alder Hey', *Modern Law Review*, Vol. 64 (2001), pp. 710–29.

Maximow, A., 'Tissue Cultures of Young Mammalian Embryos', *Contributions to Embryology*, Vol. 16 (1925), pp. 49–110.

Mayer, A.K., '"A Combative Sense of Duty": Englishness and the Scientists', in Mayer, A.K. and Lawrence, C. (eds), *Regenerating England: Science, Medicine and Culture in Interwar Britain* (Amsterdam: Rodopi Press, 2000), pp. 67–106.

Maynard-Moody, S., *The Dilemma of the Fetus: Fetal Research, Medical Progress and Moral Politics* (New York: St Martins Press, 1995).

McWhorter, J.E. and Whipple, A.O., 'The Development of the Blastoderm of the Chick *In Vitro*', *Anatomical Record*, Vol. 6 (1912), pp. 121–39.

Montgomery, P.O'B. Jr., Cook, J.E., Reynolds, R.C., Paul, J.S., Hayflick, L., Stock, D., Schulz, W.W., Kimsey, S., Thirlof, R.G., Rogers, T. and Campbell, D., 'The Response of Single Human Cells to Zero Gravity', *In Vitro*, Vol. 14, no. 2 (1978), pp. 165–73.

Morgan, L.M., '"Properly Disposed Of": A History of Embryo Disposal and the Changing Claims on Fetal Remains', *Medical Anthropology*, Vol. 21 (2002), pp. 247–74.

Morgan, L.M., *Icons of Life: A Cultural History of Human Embryos* (Berkeley and London: University of California Press, 2009).

Moscona, A., Trowell, O.A. and Willmer, E.N., 'Methods', in Willmer, E.N. (ed.), *Cells and Tissues in Culture: Methods, Biology and Physiology, Volume One* (London: Academic Press, 1965), pp. 19–86.

Mulinos, M.G., 'Cycloserine: An Antibiotic Paradox', *Antibiotics Annual* (1956), pp. 131–5.

Mulkay, M., *The Embryo Research Debate: Science and the Politics of Reproduction* (Cambridge: Cambridge University Press, 1997).

Mulnard, J.G., 'The Brussels School of Embryology', *International Journal of Developmental Biology*, Vol. 36 (1992), pp. 17–24.

Murray, D., *Aldous Huxley: An English Intellectual* (London: Little Brown, 2002).

Murray, M.R., 'Tissue Culture Procedures in Medical Installations, A: Sources and Handling of Material', in Visscher, M.B. (ed.), *Methods in Medical Research: Volume 4* (New York: Year Book Publishers, 1951), pp. 211–12.

Murray, M.R. and Kopech, G. (eds), *A Bibliography of the Research in Tissue Culture, 1884–1950: An Index to the Literature of the Living Cell Cultivated In Vitro* (New York: Academic Press, 1953).

Nardone, R.M., 'An Overview of the Scientist's Responsibilities: Comments by a Scientist', *In Vitro*, Vol. 13, no. 10 (1977), pp. 696–712.

Nathoo, A., *Hearts Exposed: Transplants and the Media in 1960s Britain* (Basingstoke: Palgrave Macmillan, 2009).

Nelkin, D., *Science as Intellectual Property: Who Controls Research?* (London: Macmillan, 1984).

Nicholson, D.J., 'Biological Atomism and Cell Theory', *Studies in the History and Philosophy of Science, Part C: Studies in the History and Philosophy of Biological and Biomedical Sciences*, Vol. 41 (2010), pp. 202–11.

O'Neill, O., 'Medical and Scientific Uses of Human Tissue', *Journal of Medical Ethics*, Vol. 22 (1996), pp. 2–5.

O'Neill, O., *Autonomy and Trust in Bioethics* (Cambridge: Cambridge University Press, 2001).

Okada, Y., 'Analysis of Giant Polynuclear Cell Formation Caused by HVJ Virus from Erlich's Ascites Tumour Cells: I. Microscopic Observation of Giant Polynuclear Cell Formation', *Experimental Cell Research*, Vol. 26 (1962), pp. 98–107.

Oppenheimer, J.M., 'Taking Things Apart and Putting them Back Together Again', *Bulletin of the History of Medicine*, Vol. 52 (1978), pp. 149–61.

Overy, R., *The Morbid Age: Britain Between the Wars* (London: Allen Lane, 2009).

Pappworth, M., *Human Guinea Pigs: Experimentation on Man* (London: Routledge and Kegan Paul, 1967).

Parry, B., *Trading the Genome: Investigating the Commodification of Bio-Information* (New York: Columbia University Press, 2004).

Parry, B. and Gere, C., 'Contested Bodies: Property Models and the Commodification of Human Biological Artefacts', *Science as Culture*, Vol. 15 (2006), pp. 139–58.

Paul, E., *Chronos, or the Future of the Family* (London: Kegan Paul, Trench, Trubner & Co., 1929).

Paul, J., 'Achievement and Challenge', in Barigozzi, C. (ed.), *Origin and Natural History of Cell Lines* (New York: Alan R. Liss, 1978), pp. 3–10.

Pauly, P.J., *Controlling Life: Jacques Loeb and the Engineering Ideal in Biology* (Oxford: Oxford University Press, 1986).

Pauly, P.J., 'Modernist Practices in American Biology', in Ross, D. (ed.), *Modernist Impulses in the Human Sciences, 1870–1930* (Baltimore: Johns Hopkins University Press, 1994), pp. 272–89.

Pauly, P.J., *Biologists and the Promise of American Life: From Meriwether Lewis to Alfred Kinsey* (Princeton: Princeton University Press, 2000).

Perley, S.N., 'From Control Over One's Body to Control Over One's Parts: Extending the Doctrine of Informed Consent', *New York University Law Review*, Vol. 67 (1992), pp. 335–65.

Pfeffer, N. and Kent, J., 'Framing Women, Framing Fetuses: How Britain Regulates Arrangements for the Collection and Use of Aborted Foetuses in Stem Cell Research and Therapies', *Biosocieties*, Vol. 2 (2007), pp. 429–47.

Pickstone, J.V., 'Bureaucracy, Liberalism and the Body in Post-Revolutionary France: Bichat's Physiology and the Paris School of Medicine', *History of Science*, Vol. 19 (1981), pp. 115–42.

Pickstone, J.V., 'Objects and Objectives: Notes on the Material Cultures of Medicine', in Lawrence, G. (ed.), *Technologies of Modern Medicine* (London: Science Museum, 1994), pp. 13–24.

Pickstone, J.V., *Ways of Knowing: A New History of Science, Technology and Medicine* (Manchester: Manchester University Press, 2000).

Pieters, T., 'Hailing a Miracle Drug: The Interferon', in de Blécourt, W. and Usborne, C. (eds), *Cultural Approaches to the History of Medicine: Mediating Medicine in Early Modern and Modern Europe* (Basingstoke: Palgrave Macmillan, 2003), pp. 142–74.

Poggi, C., 'Dreams of Metallized Flesh: Futurism and the Masculine Body', *Modernism/Modernity*, Vol. 4 (1997), pp. 19–43.

Pomerat, C., 'Use of Tissue Cultures in Drug Testing Operations', in Visscher, M.B. (ed.), *Methods in Medical Research: Volume 4* (Chicago: Year Book Publishers, 1951), pp. 266–71.

Quigley, M., 'Property: The Future of Human Tissue?', *Medical Law Review*, Vol. 17 (2009), pp. 457–66.

Rabinow, P., *Essays on the Anthropology of Reason* (Princeton: Princeton University Press, 1996).

Rabinow, P., *Making PCR: A Story of Biotechnology* (Chicago: University of Chicago Press, 1996).

Rabinow, P., *French DNA: Trouble in Purgatory* (Chicago: University of Chicago Press, 1999).

Rachels, J., 'An Overview of the Scientist's Responsibilities: Comments by an Ethicist', *In Vitro*, Vol. 13, no. 10 (1977), pp. 728–46.

Ramsey, P., *The Ethics of Fetal Research* (New Haven: Yale University Press, 1975).

Reggiani, A.H., *God's Eugenicist: Alexis Carrel and the Sociobiology of Decline* (New York and Oxford: Berghan Books, 2007).

Reynolds, A., 'The Theory of the Cell State and the Question of Autonomy in Nineteenth and Early Twentieth Century Biology', *Science in Context*, Vol. 20 (2007), pp. 71–95.

Reynolds, A., 'The Cell's Journey: From Metaphorical to Literal Factory', *Endeavour*, Vol. 31 (2007), pp. 65–70.

Reynolds, A., 'The Redoubtable Cell', *Studies in the History and Philosophy of Science, Part C: Studies in the History and Philosophy of Biological and Biomedical Sciences*, Vol. 41 (2010), pp. 194–201.

Richardson, R., 'Fearful Symmetry: Corpses for Anatomy, Organs for Transplantation?, in Younger, S., Fox, R. and O'Connell, L. (eds), *Organ Transplantation: Meanings and Realities* (Madison: University of Wisconsin Press, 1996), pp. 66–100.

Richardson, R., *Death, Dissection and the Destitute* (Second Edition: London: Phoenix Press, 2001).

Rieger, B., *Technology and the Cult of Modernity in Britain and Germany, 1890–1945* (Cambridge: Cambridge University Press, 2005).

Risen, J. and Thomas, J.L., *The Wrath of Angels: The American Abortion War* (New York: Basic Books, 1998).

Robertson, A., 'Conrad Hal Waddington. 8 November 1905–26 September 1975', *Biographical Memoirs of Fellows of the Royal Society*, Vol. 23 (1977), pp. 575–622.

Rose, H. and Rose S., *Science and Society* (London: The Penguin Press, 1969).

Rose, N., *Powers of Freedom: Reframing Political Thought* (Cambridge: Cambridge University Press, 1999).

Rosenborg, L.E., 'Using Patient Materials for Product Development: A Dean's Perspective', *Clinical Research*, Vol. 33, no. 4 (1985), pp. 452–4.

Rostand, J., *Can Man be Modified?* (London: Secker and Warburg, 1959).

Roszak, T., *The Making of a Counter Culture: Reflections on the Technocratic Society and Its Youthful Opposition* (London: Faber and Faber, 1969).

Rothman, D., *Strangers at the Bedside: A History of How Law and Bioethics Transformed Medical Decision-Making* (New York: Basic Books, 1991).

Royston, I., 'Cell Lines from Human Patients: Who Owns Them?', *Clinical Research*, Vol. 33, no. 4 (1985), pp. 422–3.

Russell, B., *Icarus, or the Future of Science* (London: Kegan Paul, Trench, Trubner & Co., 1924).

Russell, B., *Has Man a Future?* (Nottingham: Spokesman Books, 2001).

Russell, K., 'Tissue Culture – A Brief Historical Review', *Clio Medica*, Vol. 4 (1969), pp. 110–19.

Russell, W.M.S., 'The Increase of Humanity in Experimentation: Replacement, Reduction and Refinement', *Collected Papers of the Laboratory Animals Bureau*, Vol. 6 (1957), pp. 23–7.

Russell, W.M.S. and Burch, R.L., *The Principles of Humane Experimental Technique* (London: Methuen & Co., 1959).

Ryder, R.S., *Victims of Science: The Use of Animals in Research* (London: Davis-Poynter, 1975).

Sandbrook, D., *White Heat: A History of Britain in the Swinging Sixties* (London: Little, Brown, 2006).

Sanders, K., 'Tissue Cultures as Substitutes for Experimental Animals', *Collected Papers of Laboratory Animals Bureau*, Vol. 6 (1957), pp. 35–44.

Salter B., *The New Politics of Medicine* (Basingstoke: Palgrave Macmillan, 2004).

Scherer, W.F., Syverton, J.T. and Gey, G.O., 'Studies on the Propagation In Vitro of Poliomyelitis Virus', *Journal of Experimental Medicine*, Vol. 97 (1953), pp. 695–715.

Seale, C., Cavers, D. and Dixon-Woods, M., 'Commodification of Body Parts: By Medicine or by Media?', *Body and Society*, Vol. 12 (2006), pp. 25–42.

Secler, P., 'Standardizing Wounds: Alexis Carrel and the Scientific Management of Life in the First World War', *British Journal for the History of Science*, Vol. 41 (2008), pp. 73–109.

Serres, M., *Five Senses: A Philosophy of Mingled Bodies* (London: Continuum, 2008).

Shapo, M., 'Legal Responsibilities at Issue – Emphasis on Informed Consent', *In Vitro*, Vol. 13, no. 10 (1977), pp. 613–31.

Singer, P., 'All Animals are Equal', *Philosophic Exchange* (1974), pp. 103–16.

Singer, P., *Animal Liberation: Toward an End to Man's Inhumanity to Animals* (London: Cape, 1976).

Skene, L., 'Who Owns Your Body? Legal Issues in the Ownership of Bodily Material', *Trends in Molecular Medicine*, Vol. 8 (2008), pp. 48–9.

Sleigh, C., 'Plastic Body, Permanent Body: Czech Representations of Corporeality in the Early Twentieth Century', *Studies in the History and Philosophy of Science, Part C. Studies in the History and Philosophy of the Biological and Biomedical Sciences*, Vol. 40 (2009), pp. 241–55.

Smith, R., 'Biology and Values in Interwar Britain: C.S. Sherrington, Julian Huxley and the Vision of Progress', *Past and Present*, Vol. 178 (2003), pp. 210–43.

Smyth, D., *Alternatives to Animal Experiments* (London: Scholar Press in association with Research Defence Society, 1978).

Spear, F.G., 'Tissue Culture and Its Application to Radiological Research', *British Journal of Radiology*, Vol. 8 (1935), pp. 68–86.

Squier, S.M., *Babies in Bottles: Twentieth-Century Visions of Reproductive Technology* (New Brunswick: Rutgers University Press, 1994).

Squier, S.M., *Liminal Lives: Imagining the Human at the Frontiers of Biomedicine* (London: Duke University Press, 2004).

Stapledon, O., *Last and First Men* (London: Millennium Books, 2004).

Start, R.D., Brown, W., Bryant, R.J., Reed, M.W., Cross, S.S., Kent, G. and Underwood, J.C.E., 'Ownership and Uses of Human Tissue: Does the Nuffield Bioethics Report Accord with the Opinion of Surgical Inpatients?', *British Medical Journal*, Vol. 313 (1996), pp. 1366–8.

Stevens, T.M.L., *Bioethics in America: Origins and Cultural Politics* (Baltimore: Johns Hopkins University Press, 2003).

Stone, D., 'Ludovici, Anthony Mario (1882–1971)', *Oxford Dictionary of National Biography* (Oxford: Oxford University Press, online edition 2009).

Strangeways, D.E., '1905–1926', in Fell, H.B.F., Spear, F.G. and Strangeways, D.E., *The History of the Strangeways Research Laboratory: 1912–1962* (Cambridge: Heffer and Sons, 1962), pp. 7–12.

Strangeways, T.S.P., *The New Research Hospital at Cambridge* (Cambridge: Heffer and Sons, 1912).

Strangeways, T.S.P., 'Observations on the Changes Seen in Living Cells During Growth and Division', *Proceedings of the Royal Society of London: Series B, Containing Papers of a Biological Character*, Vol. 94 (1922), pp. 137–41.

Strangeways, T.S.P., 'Observations on the Formation of Bi-Nuclear Cells', *Proceedings of the Royal Society of London: Series B, Containing Papers of a Biological Nature*, Vol. 96 (1924), pp. 291–3.

Strangeways, T.S.P., *The Technique of Tissue Culture 'In Vitro'* (Cambridge: Heffer and Sons, 1924).

Strangeways, T.S.P., *Tissue Culture in Relation to Growth and Differentiation* (Cambridge: Heffer and Sons, 1924).

Strangeways, T.S.P., 'The Living Cell', *British Medical Journal* (1926), pp. 596–7.

Strangeways, T.S.P. and Fell, H.B.F., 'Experimental Studies on the Differentiation of Embryonic Tissues Growing *In Vivo* and *In Vitro*', *Proceedings of the Royal Society of London: Series B, Containing Papers of a Biological Character*, Vol. 99 (1926), pp. 340–66.

Strangeways, T.S.P. and Hopwood, F.L., 'The Effect of X-Rays Upon Mitotic Cell Division in Tissue Cultures *In Vitro*', *Proceedings of the Royal Society of London: Series B, Containing Papers of a Biological Character*, Vol. 100 (1926), pp. 283–93.

Strangeways, T.S.P. and Oakley, H.E.H., 'The Immediate Changes Observed in Tissue Cells After Exposure to Soft X-Rays while Growing in Vitro', *Proceedings of the Royal Society of London. Series B, Containing Papers of a Biological Character*, Vol. 95 (1923), pp. 373–81.

Sun, M., 'Scientists Settle Cell Line Dispute', *Science*, Vol. 220 (1983), pp. 393–4.

Tallerico, C.A., 'The Autonomy of the Human Body in the Age of Bio-technology', *University of Colorado Law Review*, Vol. 61 (1990), pp. 659–80.

Taylor, G.R., *The Biological Time-Bomb* (London: Thames and Hudson, 1968).

Thomson, D., 'Controlled Growth en masse (somatic growth) of Embryonic Chicken Tissue In Vitro', *Proceedings of the Royal Society of Medicine: Laboratory Reports*, Vol. 7 (1913), pp. 71–5.

Thomson, D., 'Some Further Remarks on the Cultivation of Tissues *in vitro*', *Proceedings of the Royal Society of Medicine*, Vol. 7 (1914), pp. 2–46.

Thomson, D. and Thomson, J.G., 'The Cultivation of Human Tumour Tissue *in Vitro*', *Proceedings of the Royal Society of Medicine, London*, Vol. 7, no. 1 (1913–1914), pp. 7–20.

Thomson, D. and Thomson, J.G., 'The Cultivation of Human Tumour Tissue *in Vitro* – Preliminary Note', *Proceedings of the Royal Society of Science: Series B, Containing Papers of a Biological Character*, Vol. 88 (1914), pp. 90–1.

Titmuss, R., *The Gift Relationship: From Human Blood to Social Policy* (London: Allen and Unwin, 1971).

Tudor, A., *Monsters and Mad Scientists: A Cultural History of the Horror Movie* (Oxford: Basil Blackwell, 1989).

Turney, J., *Frankenstein's Footsteps: Science, Genetics and Popular Culture* (London: Yale University Press, 1998).

Tutton, R., 'Person, Property and Gift: Exploring Languages of Tissue Donation to Biomedical Research', in Tutton, R. and Corrigan, O. (eds), *Genetic Databases: Socio-ethical Issues in the Collection and Use of DNA* (London: Routledge, 2004), pp. 19–39.

United Kingdom Coordinating Committee on Cancer Research, 'UKCCCR Guidelines for the Use of Cell Lines in Cancer Research', *British Journal of Cancer*, Vol. 82, no. 9 (2000), pp. 1495–509.

Vacanti, C., 'The History of Tissue Engineering', *Journal of Cellular and Molecular Medicine*, Vol. 10 (2006), pp. 569–76.

Vaughan, J., 'Honor Bridget Fell: 22 May 1900–22 April 1986', *Biographical Memoirs of Fellows of the Royal Society*, Vol. 33 (1987), pp. 237–59.

Vogel, G., 'FDA Weighs Using Tumour Cell Lines for Vaccine Development', *Science*, Vol. 285 (1999), pp. 1826–7.

Vyvyan, J., *In Pity and in Anger: A Study of the Use of Animals in Science* (London: Joseph Press, 1969).

Waddington, C.H., 'Induction by Coagulated Organisers in the Chick Embryo', *Nature*, Vol. 131 (1933), pp. 275–6.

Waddington, C.H., 'Experiments on the Development of Chick and Duck Embryos, Cultivated *In Vitro*', *Philosophical Transactions of the Royal Society of London. Series B, Containing Papers of a Biological Nature*, Vol. 221 (1932), pp. 179–230.

Waddington, C.H. and Waterman, A.J., 'The Development *In Vitro* of Young Rabbit Embryos', *Anatomy*, Vol. 57 (1932), pp. 355–70.

Wade, N., 'Hayflick's Tragedy: The Rise and Fall of a Human Cell Line', *Science*, Vol. 192 (1976), pp. 125–7.

Wade, N., 'University and Drug Firm Battle Over Billion-Dollar Gene', *Science*, Vol. 209 (1980), pp. 1492–4.

Wagner, A.B., 'The Legal Impact of Patient Materials Used for Product Development in the Biomedical Industry', *Clinical Research*, Vol. 33, no. 4 (1985), pp. 444–7.

Waldby, C., *The Visible Human Project: Informatic Bodies and Posthuman Medicine* (London: Routledge, 2000).

Waldby, C. and Mitchell, R., *Tissue Economies: Blood, Organs and Cell Lines in Late Capitalism* (Durham and London: Duke University Press, 2006).

Walton, A.J., 'The Effect of Various Tissue Extracts Upon the Growth of Adult Mammalian Cells In Vitro', *Journal of Experimental Medicine*, Vol. 20 (1914), pp. 554–72.

Waymouth, C., 'Construction and Use of Synthetic Media', in Willmer, E.N. (ed.), *Cells and Tissues in Culture: Methods, Biology and Physiology, Volume One* (London: Academic Press, 1965), pp. 99–132.

Weatherhall, D.J., '*Daedalus*, Haldane and Medical Science', in Dronamrjau, K. (ed.), *Haldane's Daedalus Revisited* (Oxford: Oxford University Press, 1995), pp. 102–24.

Weatherhall, M., *Gentlemen, Scientists and Doctors: Medicine at Cambridge, 1800–1920* (Cambridge: Boydell Press, 2001).

Wells, H.G., Wells, G.P. and Huxley, J., *The Science of Life* (London: Cassell and Co., 1938).

Werskey, G., *The Visible College: A Collective Biography of British Scientists and Socialists During the 1930s* (London: Allen Lane, 1978).

White, P.R., *The Cultivation of Animal and Plant Cells* (New York: The Ronald Press Co., 1962).

Willmer, E.N., 'Studies on the Influence of the Surrounding Medium on the Activity of Cells in Tissue Culture', *British Journal of Experimental Biology*, Vol. 4 (1927), p. 280.

Willmer, E.N., 'Tissue Culture from the Standpoint of General Physiology', *Biological Reviews of the Cambridge Philosophical Society*, Vol. 3 (1928), pp. 271–302.

Willmer, E.N., *Tissue Culture* (First Edition. London: Methuen & Co., 1935).

Willmer, E.N., *Tissue Culture* (Third Edition. London: Methuen & Co., 1964).

Willmer, E.N., 'Introduction', in Willmer, E.N. (ed.), *Cells and Tissues in Culture: Methods, Biology and Physiology, Volume One* (London: Academic Press, 1965), pp. 1–17.

Wilson, D., 'Historical Keyword: "Tissue"', the *Lancet*, Vol. 373 (2009) p. 109.

Winslade, W.J., 'An Overview of the Scientist's Responsibilities: Comments by an Attorney', *In Vitro*, Vol. 13, no. 10 (1977), pp. 712–27.

Witkowski, J.A., 'Alexis Carrel and the Mysticism of Tissue Culture', *Medical History*, Vol. 23 (1979), pp. 279–96.

Witkowski, J.A., 'Dr. Carrel's Immortal Cells', *Medical History*, Vol. 24 (1980), pp. 129–42.

Witkowski, J.A., 'Honor Fell', *Trends in Biochemical Sciences*, Vol. 11 (1986), pp. 486–8.

Witkowski, J.A., 'Ross Harrison and the Experimental Analysis of Nerve Growth: The Origins of Tissue Culture', in Horder, T.J., Witkowski, J.A. and Wylie, C.C. (eds), *A History of Embryology: The Eighth Symposium of the British Society for Developmental Biology* (Cambridge: Cambridge University Press, 1986), pp. 149–77.

Index